581.1

D1324467

Roots: Miracles Below

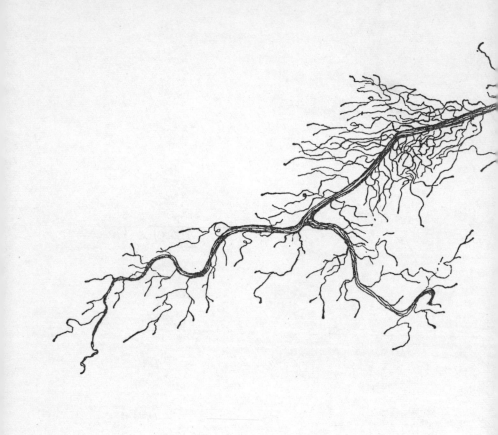

26897

Roots: Miracles Below

Charles Morrow Wilson

DOUBLEDAY & COMPANY, INC.   GARDEN CITY, NEW YORK

Illustrations drawn by Grambs Miller

Library of Congress Catalog Card Number 68–17806

Contents

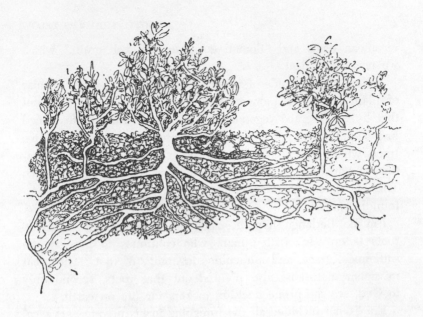

MIRACLES BENEATH OUR FEET

Plant roots are the silent, little-seen, uncountably numerous miracle makers and life sustainers.

Earlier generations of natural scientists defined roots as the underground digestive tracts of vegetation. One such definer was a first agricultural missionary to the New World, Fra Tomàs de Berlanga of Grand Canary Island. Back in 1419 Fra Tomàs began importing and planting the roots and seeds of already established food crops from the Canary Islands to help safeguard Hispaniola, or Dominica, Panama, and other Caribbean lands against recurring famines. Crop historians credit the friar, who became the first bishop of Panama, with having introduced bananas and plantains to the New World. There is evidence that the cleric friend of Christopher Columbus also introduced other people-feeding crops, including oats from the lower Medi-

terranean lands and "Roman corn" or "English grain," which
we now call wheat.

Among his multitude of talents Brother Thomas was a scholar
of plant life. As a pioneer plant physiologist he pointed out
that by God's grace vegetation is essentially animal life turned
inside out. Leaves are its lungs. Roots are its unencased digestive
tracts which largely sustain themselves and nurture the vegetation of which they are part. Sunlight is the energizing heart.
The circulatory systems of vegetation and animals are profoundly similar.

Fra de Berlanga hypothesized that God saw fit to create
roots before He created man, who continues to live by the
sufferance, grace, and miraculous ingenuity of roots. The great
missionary-naturalist also pointed out that roots, second only
to God, are the prime decision makers for life on earth.

Fra Tomàs included all the foregoing in his private encyclical
of faith. Plant scholars today tend to list his tenet as science—
from *scire—scientis—*"to know." Among the growing throng of
present-day scholars of roots is one Johannes van Overbeek, a
long-time teacher of plant physiology at the University of Utrecht
until the Carnegie Foundation brought him to the United States
to "help arouse teacher interest in his subject." Subsequently
he joined the staff at California Institute of Technology, then
went into plant research for the U. S. Government, and presently
is in major-league commercial research, and has always been a
diligent student of roots.

Dr. van Overbeek sustains Fra Berlanga's conviction that
roots are primary decision makers; as a teaching scientist he
developed a fairly simple experiment to support that thesis.
The experiment begins by stem grafting two commonplace
and reasonably hardy plant kinds—tobacco and tomato. The
feat is not particularly difficult or complex: one simply grafts
the above-ground portion of a young tomato plant to a root
crown or base stem of a like-sized tobacco plant.

When the graft "takes," the resulting plant will continue to have stem and leaf growth that look fairly much like those of any other tomato plant. The resulting plant will not bear tomatoes. But its leafage will contain that ever-alluring alkaloid ($C_{10}H_{14}N_2$) nicotine. Press the partly dried leaves and the telltale acrid, oily, almost colorless fluid that is the ever-special hallmark of the genus *Nicotiana* (just call it tobacco) will presently ooze out. The decision-making roots will have seen to that.

Call the vegetative mixup tobacco-mato or tomato-bacco or whatever you like, the above-ground portions of the man-contrived merger when duly processed is smokable, chewable, or sniffable as tobacco. It will befoul the breath, stain the teeth, and pollute the bloodstream like any other of its devilishly attractive kind. True, it may not be so mild as the first gentle kiss of spring or the first subtle brush of Madison Avenue horse feathers, but it is consumable as tobacco; those decision-making roots have got in their engaging skulduggery.

But there is no *vice versa* factor. Stem-graft the above-ground portion of a tobacco plant to the head of a tomato root system, plant it, watch it grow, harvest and cure it as you will, then roll it into cigarettes or cigars, stuff it into your pipe, or chew it, God forbid, and you will taste, smell, or otherwise find or feel no trace whatsoever of nicotine; not one tobaccoish wisp in a carload. Again and as usual, the roots have dominated and decided—in this instance against nicotine, the outrageously alluring hallmark of tobacco.

Johannes van Overbeek makes it clear that he does not conspire to bring about miscegenation between tobacco and tomatoes or any other species. As a research scientist and teacher he holds the absolute conviction that roots are indeed the premier decision makers, providers, and more gentle defenders of life on land. Furthermore as a practical scholar he constructively practices what he learns and teaches about roots. For example, in a long-shunned wasteland near his Modesto, California, home, he has established and brought to fruition—

all from roots, he emphasizes—a magnificent vineyard of wine-flavored grapes. Country neighbors, who had joined in nick-naming the unwatered thin-soiled terrestrial junkyard, so to speak, Starvation Rise, still find it hard to believe. They laughed and laughed when the hard-working, quietly observant "perfes-ser" set to planting the dismal little no man's land with rooted cuttings and devising a Rube Goldbergish watering and drain-age system. Not long thereafter the tenor of the neighborly comments changed drastically. "What we're concerned about now," one of the neighbors recently told me, "is that our root-tinkering professor friend is getting more grapes than he can clip and press or even count out. If this goes on—this hand-up-by-grape-roots, I mean—man, we could all get washed hellwards by flash floods of van Overbeek grape juice."

The concerned neighbor added that the root magician ap-pears to sort of fade out when it comes to defining in plain, understandable American precisely what he means by "root power" or, for that matter, even "roots."

In this matter of communication gap the distinguished plant physiologist has plenty of company, including word-juggling company. For example, the writer has been checking through the going crop of dictionary and encyclopedia definitions of "root." The agreed-on point that the English "root" is from the Anglo-Saxon "rot" may be symbolic, at least to a dour degree, but even as a first step in definition it is a stumbler. Much the same applies to every definition I can locate. At this point of the search it seems to me that the newest (1961) Funk & Wagnalls two-volume desk dictionary offers the most cogent of the nineteen efforts at defining "root" that I have thus far tracked down, picked up, and tried to weigh. By all that is true and proper, I am not an habituated dictionary heckler. As a onetime encyclopedia writer I like dictionaries and feel tolerant toward their quaint, antiquarian frailties and senile vagaries. Furthermore, I would be the last to lambast the products of the onetime Messrs. Funk and Wagnalls. Actually, I rate theirs as one of the better of the abridged or so-called desk dictionaries now available, and repeat that compared with the rest, it seems to me to make the best or certainly one of the better bites at briefly defining "root":

"Root: 1. The underground portion or descending axis of a plant which absorbs moisture, obtains or stores nourishment, and provides support. It differs from the stem in that it branches irregularly and lacks joints or leaves. 2. Popularly, any underground growth, as a tuber or bulb."

But the best is simply nowhere near good enough. To pick a convenient starting point: The root of a plant is not necessarily or invariably the underground portion. There are root systems, such as those of *Hydrocharis* ("frog's-bit" is one of its better known species), which live and function entirely in water, and others, including the *Angraecum* orchids that grow entirely in the air and for fuller measure function like leaves. There are also parasitic roots, such as mistletoe (*Phoradendron*) and broomrape (*Orobanchae*), that are nowhere near underground, but live from the sap streams of other plants and therefore from underground roots only by remote control. And as any southbound tourist can learn from fleeting glimpses, there are also plants that live and multiply without any true roots. The most familiar is Spanish moss (*Tillandsia usneoides*), a plant doubly misnamed—it is neither a moss nor Spanish, but rather is a bromeliaceous epiphytic (no hard feelings), one of the many oddball kin of the pineapple family. As a particularly expert feeder on air, the *Tillandsia* can live as easily on a telephone wire as on a tree branch and would have nothing to do with an "underground portion."

Furthermore, a root is not necessarily a "descending axis." Roots of swamp cypresses, to cite just one example, raise erect "knees" into the air. Many plant species, including some of the tree ferns, have root structures that grow upward instead of downward in order to encase and presumably protect their lower stems. Greater numbers of plants have roots that grow laterally and, in many instances, branch out from above-ground stems. One of the more readily noticeable examples is common field corn or maize, now the most widely grown row or field crop of the Western Hemisphere. As practically everybody will recall, corn also has above-ground roots which are attached like guy wires to the lower joints of the stalk. The screwpine (*Pandanus*) and the banyan (*Ficus benghalensis*) are among the fairly numerous species of trees with roots that regularly grow above ground.

The statement that roots do not have joints or leaves is also subject to a great many exceptions. To list only a few, the surviving descendants of the ancient race of plants called horsetails still have root systems made up partially of jointed hollow stems. What's more, many of the so-called higher-order plants—plum and cherry trees for ready examples—more or less regularly sprout leaves directly from their surface roots. By return courtesy, so it would seem, thousands of plant species have stems that can and do produce roots with the greatest of ease, either with or without encouragement by man. Indeed, when stems are clipped from almost any woody or semiwoody plant and placed directly into damp soil or sand, miniature roots are likely to appear, even though some cannot survive as productive roots. The plant physiologists tell us these stem roots—on cuttings —develop from microscopic cells called root primordia. Rose bushes and the common fruit trees are among the most frequently exploited examples. For very good reasons, commercial nurseries and professional plant breeders are relying more and more heavily on plant propagation by means of cuttings, which, in many instances, uphold species characteristics far better than seeding and provide much hardier planting stock.

As anybody who bothers to look knows, roots also grow from stem surfaces or branch tips that come into contact with soil, as with loganberries; or from leaves that make contact with the soil, as with the Rex begonias. Other plants, such as strawberries, have above-ground runners which take root and produce new plants at intervals. Alfalfa, in its own deft way, spreads in similar fashion underground. Recently I exhumed fourteen alfalfa plants in a single colony, all rooted independently but connected by means of strands, living connectors, a few inches below the surface.

The secondary definition—"Popularly, any underground growth, as a tuber or bulb"—also invites argument. Many people, including plant physiologists and botanists, would stoutly deny that all underground "growths" are roots. A bulb consists mostly of enlarged and otherwise modified leaves, with roots attached to the basal plate, which, so botany texts assert, is the true stem. A rhizome is an enlarged underground stem. In many instances it sends up leafy shoots from the upper surface and sprouts roots below. A tuber, of course, is not literally a root either, but the fact that it lives in root spheres (the going shoptalk word is "rhizospheres") and is principally fed by roots establishes a vital affiliation or, as some plant students prefer, "annona" with true roots. But technical definitions aside, bulbs, rhizomes, and tubers are all inhabitants of the rhizospheres. Rhizospheres in themselves are usually very special subsurface worlds in which vast numbers of much smaller plants and animals also live and compete for livelihoods, including space, moisture, and nutrition—no holds barred and no arbitrary segregation attainable.

There is extremely good reason to believe that this has been going on for hundreds of millions of years longer than man has lived on earth. A most powerful confirming evidence, and perhaps the most telling testimony to the superlative importance of roots, is part and parcel of the routine processes of plant germination.

Usually when a seed germinates, the primary roots, or radicle, is the very first part of the borning plant to emerge. Usually the tiny, pallid taproot penetrates its rhizosphere or soil area with valiant determination, and given any kind of a chance, sustains its plant with magnificent loyalty and vigor. Once in creative action, the living root structure keeps on writing and living its own definitions.

This again brings home the dogged refusal of roots to keep with any absolute definition, however cogent, or any absolute answers, even for such entirely simple questions as, How deep do roots grow? and, How far do they spread? There aren't any really definitive rules. In a great many instances the root depths of a given plant are at least equal to the above-ground height of that plant. We know that some of the most valuable and longest-lived grasses have a handful or hatful of top and virtually a barrelful of roots. We know, too, that a hardy alfalfa plant which is a foot high and, say, seven or eight years old may well have a lifeline of roots as much as thirty feet long. Even in the instance of an annual, single-season grass, such as corn or maize, the primary root system may penetrate the soil to depths of anywhere between four and eight feet. However, if the root system of a hardy four-month-old corn plant were elongated as a continuous living strand, the total line of functioning tissues would very probably be somewhere between 400 and 600 feet in length—for just this single corn plant. Thus, for a ten-acre field, total underground root length might very well total at least 5000 miles. Root structures of big and long-lived forest trees are known to reach downward as far as 100 feet. Roots of the famed California redwoods (*Sequoia sempervirens*) are known to penetrate subsoil areas of up to 50,000 cubic feet. If one attempted measuring or even making educated guesses about the total root lengths in a medium-sized county of farms, he would almost certainly have to begin the count in quadrillions of feet, which is about as incomprehensible as logging mileages of outer space.

Measuring the actual reach of roots relates to the facts that until a root system dies, it cannot cease growing, and when a root system dies, the entire plant structure almost invariably dies. On the paradoxical side, roots must keep growing, too, because they are continuously dying.

In any case, and stated quite certainly, the size of a root structure does not necessarily decide its strength or vitality. Certainly one should not be deceived by the delicate appearance of the first thread dimension roots of a plant or by the lacelike traceries that in many instances characterize the network of a mature root system. The fragile-looking tips or caps of very tiny rootlets can thrust aside practically all smaller objects that obstruct them and can pierce hard, compacted soil. The sheer power of roots is practically immeasurable. For one readily observable example, when tree roots wedge themselves into the crevices of a granite rock, it is most likely to be the rock that ultimately yields, not the roots. The fact usually obtains that the granite is split asunder and the roots, which only *look* soft and fragile, continue on their way, growing, feeding, refining, and dispatching foods to their plants even while bearing the vast weight of the rock above them.

In terms of age, roots are unquestionably among the oldest continuing life forms on or in earth; many scholars contend the very oldest. Some years ago H. W. Staten, a distinguished American agronomist who climaxed a long teaching career by serving as agricultural advisor to Ethiopia and Emperor Haile Selassie, came forward with the carefully studied estimate that the root structures of big bluestem, a perennial forage grass believed native to the American Great Plains, remain alive for at least 6000 years. The bluish-green tops of big bluestem sustain the name, but clearly the immense root structure which sustains the species is the immortality factor. Dr. Staten therefore submitted that roots of big bluestem could very well be the "oldest living things now on earth." He agreed with the paleobotanists, ecologists, and other scholars that bristlecone pines, which still thrive in the U. S. Far West and elsewhere west of the hundredth meridian, have lived from the same ever-growing root structures for no less than 4000 years, and that the roots of some of the famed sequoias or redwoods of northern California may well have been thriving during the time of Aristotle, three to four centuries before the time of Christ.

Here in the United States, particularly in Florida and other Southeastern areas, there is evidence of other root lives that apparently extend back through more centuries than anyone can surely count. Among leading contenders here are the cycads. Some species of these vegetative ancients are believed to have roots that must live for at least 1000 years before their plants even begin to produce seeds. Ordinarily one does not associate roots with cycads, which one usually sees merely as rosettes of dark green spikes which seem either to attach themselves haphazardly to lawns or other open surfaces or to hide away in the wilds. But the usual fact is that whether they hug the ground or raise cyclindrical stems to heights of as much as twenty feet, their living tissues under and directly above the soil surface far exceed in weight that of their combined leaf and stem structures. In many species the so-called base stands are distinctly tuberous—sizable swells of tissue which descend into fleshy taproots and very extensive feeder roots.

Many species of the cycads grow almost completely underground, showing only very tiny leaf tips or seed cones above ground. The manner of growth of the cycads suggests the past and quite possibly the future of great underground food crops. In some lands, including Madagascar and Australia, the leaves, seeds, stems, and root heads of some kinds of cycads are used extensively for human food. Cycad families, such as the Australian *Bowenia*, have heavy underground stems similar to giant carrots, and are validly edible.

From time to time, like lightning flashes in darkness, new revelations of the almost fabulous durability of roots come thrusting upon us. One fairly recent instance was the astonishing rediscovery of dawn redwood, one of the so-called living fossils. Supposedly, this coniferous ancestor of the California sequoia completed its earthly span millions of years ago. Yet during

1941, reports came that the allegedly long-extinct tree ancestor (which had been punished posthumously by the somewhat horrible name of *Mestaseopia Glyhptostroboide*) is still living and flourishing in various areas of lower interior China. During 1947 a botanical expedition from Harvard succeeded in bringing back viable seed of the supposedly extinct conifer. In the Arnold Arboretum near Cambridge, the seed germinated and flourished. More recently the U. S. Department of Agriculture's plant introduction station in Glendale, Maryland, joined in welcoming back to life the misnamed "living fossil" and providing it a more widespread state of being.

In its general appearance the dawn redwood resembles the bald cypress. It usually grows to a height of about seventy feet, and frequently reaches thirty-five feet during its first ten years. The above-ground silhouette is conical. In autumn and winter the bright emerald foliage turns coppery red. The very attractive tree endures subzero temperatures and appears to have no chronic diseases or insect enemies. What has particularly interested me is its root structure, which is compact and very well developed. It has a long, centrally placed tap root, supplementing branching roots, and a lusty and very extensive system of feeder roots equipped with hard, clearly formed drilling caps and microscopically small root hairs; like most other root hairs, these are sprangling, elongated fluid tubes apparently formed from single tissue cells. What I find most impressive is that here is a tree life which was thriving before man came on earth and is still thriving. One can reasonably infer that its superbly developed roots were superbly developed millions of years ago.

While watching its transplanting, I reflected on the ancient yet insistently modern miracle of these roots, which to mere transient man are practically immortal. In their realms of partial darkness the successive root structures of this dawn redwood have survived millions and millions of winters and summers, ranging from steamy tropical heat to frigid, earth-stiffening cold.

Like those of thousands of other plant species, its uncountable successions of root lives have worked around clocks and calendars and all other measurers of time. They have sustained uncountable generations of living trees, probably with individual lives measured in thousands of years, all the while feeding, drinking, and breathing for themselves.

At the transplantings, a solemn young attendant from the Harvard forestry group spoke knowingly of the exacting air requirements of these and other ground roots. He pointed out that the root structures in general must "breathe" extremely wet air in the soil or subsoil, with humidity ranging narrowly from 98.81 per cent of saturation to 99.3 per cent. This all-but-immortal root structure continues to fulfill the multitude of exacting requirements of present-day functions, thereby remaining as "modern" as tomorrow morning.

While viewing the contemporary ancient, I reflected on how the presence of roots materializes from the absence of roots. It is rather generally agreed that the beginning of earthly life was marked by vast legions of rootless plants, ancestors of vegetation as we now know it. Great numbers of these rootless plants still live. They include the single-celled flora called bacteria, the yeasts, many of the algae, and some of the fungi. It is at least a good guess that a majority of these rootless plants that survive are single-celled life forms, but it is also common knowledge that we have more highly developed vegetation that even now lacks identifiable roots, stems, or leaves. This emphasizes the changing forms and functions of roots as indicated by the ever-advancing or changing ways of plant life as we now see it. Any attempt, however modest, to answer the question, What are roots? invites at least a passing glance at what are conveniently dubbed primitive roots, many of them quite easy to find.

Among the better demonstrations of the continuing changeovers from rootless to rooted plants are the liverworts and the mosses. The first-named are small and generally inconspicuous.

If we notice the liverworts at all, we are most likely to see them as little, flat green objects with hornlike projections. They are amazingly self-sufficient (possibly for this reason, their race has endured through the ages) bafflingly apart from better-known realms of vegetation. Some, but by no means all, of the petite liverworts have roots, but even these are not the kinds usually recognized and classified as such. On the lower surface of their flat, leaflike thallus one finds small hairs or rhizoids which serve effectively as anchors. Apparently, at least as yet, they do not function extensively or adequately as roots.

The mosses go a step further. In their apparently primordial —or wouldn't it be better just to say first in order of time?— leaves, as well as in their rootlike or hairlike rhizoids, they show impressive similarities to the higher or better-developed plants. By means of these elementary roots, mosses take water and chemical salts directly from soil, as well as from rocks, decaying wood, and various other surfaces. They cannot grow extensively without water, but they can live for remarkable, perhaps unprecedented lengths of time without it. This is one good reason why one of the greater services of the mosses is to heal the scars of the earth. Colonizing on rocks and rock debris, on dead plants, or on burned areas, they definitely help restore the living soil; with their tiny minimal roots they clearly aid in sustaining and rebuilding the domain for more elaborate roots. In the mosses, too, the true stem emerges. The flat appendages that serve as breathing devices are rudimentary leaves. From the stems, particularly those that are in contact with soil, grow tubular and branching hairs which are incipient roots.

The stems, leaves, and roots work together to provide the moss plant with moisture and nutrients. In the center of the stem is an area for storing food. Quite clearly these formative or evolving structures point the way toward the highly specialized roots of the flowering plants.

Apparently, aerial roots were much more common in earlier

geologic times than they are today. If one looks at a common horsetail—that curious jointed, hollow-stemmed relic of the past which is so frequently findable close to one's home—one sees at the base of each stem sector an incipient aerial root. Normally the aerial roots are now dormant. The stem of the very ancient survivor grows up from the working root, which, like the stem, is principally a segmented tube, but one well equipped with fibrous clusters of feeder roots. Other aspects of the horsetail— the large air spaces, the minute leaves, and the vascular tissues —suggest that its remote ancestors grew entirely in water.

The lycopods, or club mosses, are among other survivors of ancient eras. Nowadays we know them by such rather senseless names as ground cedar, running pine, or ground pine. They are tiny in comparison with their giant ancestors of the coal-forming times; most of the present-day descendants are only a few inches high. The earthly span of the club mosses is expertly guessed at as a quarter of a billion years. It is commonly assumed that the original members of this family became extinct many millions of years ago. Apparently some of the midget forms have survived without extensive change. Sprangled, threadlike roots continue to anchor them even as in ages past, and to collect water and mineral food from the soil.

The prime structure of the present-day ferns is a basic stem, seldom visible except in the tree ferns, which bears distinctively shaped leaves called fronds. These vary in size, along with the plant, from a fraction of an inch to scores of feet. In any case, the fern stem is tethered to the soil by a clearly defined root system which carries dissolved minerals through the vascular strands—the pipelines of the plant—into the very tips of the fronds. This development, so botanists tell us, clearly marks the genesis of what they call the higher plants.

The most valuable surviving (and flourishing) race or, as some say, tribe of very ancient plants is the conifers, a name designating the seed cones which most of the member species continue to bear. There is superb evidence that the conifers have lived on earth for tens of millions of years; at least 550 tree-sized species are still with us. By totals they comprise a principal part of the remaining forests, and supply the greater part of our commercial lumber, much of our paper and other pulp-wood products. Many scholars continue to regard the conifers as evolutionary way markers for the even more provident orders of vegetation.

The point here is that plant physiologists are generally agreed that the most remarkable and life-prolonging resource of the conifers is their root systems, which many scholars insist are the most effective yet developed. Unquestionably, conifer roots are among the hardiest and most disease-resistant of all. For amazing good measure, they have the ability, still not fully understood, to collect inorganic nitrogen from the soil and convert it into the all-sustaining plant and animal food factor—organic nitrogen or protein. Thus, in terms of feeding people, this amazing ingenuity of conifer roots could well be the most important secret on earth.

Since flowering plants first appeared on the earth (according to the prevailing interpretation of fossil records, this was during the Mesozoic era, some 130 million years ago) increasingly lovely and useful plant forms have been spreading across the earth. One type makes its debut when the sprouting seeds send up a single leaf; another begins by sprouting twin leaves. Thus, the two most basic names or labels in botany are "monocots" (short for monocotyledons, the one-seed-leaf plants) and "dicots" (for dicotyledons, those that first display a pair of leaves).

The monocots include our true grasses (*Gramineae*), which, in turn, include all our principal food grains, and the largest race of flowering plants. Lilies, irises, onions, pondweeds, palms, arums, orchids, and bananas are among other members. Their stems, even when woody, are usually without age rings or other absolute time marks. But for our special interest here the monocots show in evolutionary detail the entire range of root structures, including water roots, air roots, parasitic roots,

air-and-soil roots, and the especially magnificent deep, matted
roots of the perennial grasses; all in all, they do an excellent
job of filling in the score sheet of root types.

As a vast group, the monocots accentuate the importance of
root systems. The dicots stress their vital functions even more
notably. The dicots number in their ranks the great majority
of plants that grace the so-called flowering kingdom. If you
miss seeing their twin leaves push up from the earth, you can
recognize them by other characteristics. Their leaves are almost
always elaborately veined and their flower parts usually come
in twos, fours, or fives. The stems of most have a well-defined
(though sometimes microscopic) cambium or growth layer (lacking in the monocots), and the vascular tissues are arranged in
a cylinder.

Again excepting the grains, most of our principal agricultural
crops—all the leading vegetative food crops of the United States
and practically all other temperate-zone places—are dicots; as
are the majority of forest and shade trees and most of our
flowers and flowering shrubs. The group includes the buttercup,
water-lily, and magnolia families, the rose family, the legumes
(peas and beans), the poppy and mustard families, the heaths
(heather, blueberry, rhododendron, etc.), the potato, mint, and
snapdragon families, the sunflower and its many relatives, and
great lumber and forest trees such as oaks, beeches, birches,
hickories, and ash.

All in all, the interrelations among roots, stems, and leaves
are immensely revealing and growing more so. This, of course,
makes the undersurface realms more challenging to all students,
regardless of their practical or scientific goals. The dicots are
especially revealing. With growth, the stems develop concentric
rings through which dissolved nutrients gathered by the roots
are distributed, and self-manufactured products are carried from
the leaves to all other parts of the plant. In most perennial
dicots, the stems also serve as recording calendars; each year's
growth adds a ring of wood. An especially valuable lesson (and

hallmark) of this most advanced of the vegetative orders relates to the truly wonderful interrelationship among root growth, stem growth, and leaf growth. Apparently in all instances growth begins with one meristem (cell-multiplication area) in the root and another meristem in the stem. Apparently, too, both are able to multiply virtually without limit.

As the above-ground portion of the dicotyledon emerges, its branching structure somehow acquires meristems, or cell nurseries, at every tip and bud. In the root system an even more dynamic procedure of multiplication is in progress, to balance and sustain the increasing hundreds and thousands of above-ground meristems. This seems to be the fundamental factor and controller or governor of balanced growth—with roots the prime providers, shapers, starters, and sustainers. All makes for a particularly engaging lesson, straight from the living book of Mother Nature. The first point of the lesson is that the same wonderful living complex that does most to sustain vegetation also does most to sustain the animals, including man. Without roots we simply wouldn't be here. As one continues to study roots this gets to be as self-evident as an old-fashioned country wash day.

2

ROOTS ARE ALSO FOR EATING

"Roots serve to perpetuate, sustain, restore, and resurrect. They are the most productive tenants and the most able conservers of the living film called soil. They are the all-provident joiners of past, present, and future."

My host and patient listener smiled tolerantly, then explained that he was a practical man—as a Vai tribe chief living and leading in the lower Cavally Valley of West Equatorial Africa has to be a practical man, not a conservational essayist. Paramount Chief and District Commissioner Fromo Abuyi therefore got to the kernel: "Roots are for eating."

The handsome Vai was speaking from firsthand experience and ancient and people-saving precedence. So his immediate reference was to a famine that was reportedly spreading through extensive areas of the West African coastlands from hinterland

Liberia well into the Cameroons. As an employee of an American rubber company I had been sent down to find out what it was all about.

Fromo Abuyi was briefing me with gracious competence. A very damaging drought had settled in, having apparently ridden the harmattans, the searing dry winds from the Sahara. The "big dry" had dealt serious damage to the rice crop, which is mostly so-called red rice, grown without benefit of irrigation. Not for anyone's good, some very exceptional elephant marauding from rogue herds, which only occasionally forage in that area, had lately made the bad situation worse. The latest rice crop was almost a complete failure; the local hunting was extremely poor; the fishing never was worth a whoop or a holler so far as anybody remembers.

The materializing famine was spreading worrisomely. But the local chief (who holds a degree in the humanities from the University of Beirut) explained that he did not seek or wish any outside help, including what is known as foreign aid from the United States, Soviet Russia, or anyone else.

"My people are beginning to feel hunger, but they will not starve nor suffer the pains or the diseases that come of empty bellies. The kinds of food that your country or your company would send here would not, in any case, suit my people. Ordinarily we eat rice instead of wheat or corn, palm oil instead of butter, plantains instead of potatoes, and forgive the expression, the hell with your dehydrated beets and carrots.

"I grant that your American foods have merits, as does your willingness to share them with us," he conceded. "I grant that most of this great talk about 'traditional eating' is, as you Americans say, hogwash. Your American foods are now being sold and served almost everywhere I travel, and in your cities I find your people waiting in line to eat what are called Chinese, Hungarian, Armenian, Italian, French, Middle Eastern, or whatever foods. Granting all this, and granting your American way of dispatching food a third or even half of the way around the earth is too wasteful to be sensible, food demands are getting to be more and more internationalized. What I would say here, is that God, in His wisdom, granted that every part of the earth with people has its own good and worthy food sources. What we must all learn is to eat more of what is both nourishing and close at hand." The young chief repeated with an eloquent gesture, "Now and for the lean months ahead my people will not starve. We will live through this famine time on roots, as we have done so many times before. Such is our very special light and resurrection."

He added that he had already turned to edible roots for the duration of the lean period and was finding them good. He presently demonstrated by serving me what he termed a native-root "country chop"; as any African traveler knows, that means the bowl of boiled grain topped with any available meat, fish, fruit, berries, or nuts to complete the prevailing one daily meal, usually eaten in the afternoon. The usual ingredients for Vai-style country chop are boiled rice and "small beef" (meaning any kind of available meat from snake to elephant), topped with peanuts, pineapple, or native citrus or palm fruit, for the most part picked from the wilds.

Fromo's root chop was readily edible—short of being the best country chop I have ever eaten, but very far from the worst. I remember from Fromo's creation such engaging flavors as those from licorice and ginger roots presumably wild-growing; also a whitish taproot that is flavored like yam or sweet potatoes, and a dry, mealy root probably of the sunflower family, tasting like Jerusalem artichoke. The topping or "spice curry" was made of sweet, wild-growing pineapple with a sauce of palm oil. I remember vividly how the various ingredients merged into a rather excitingly flavored composite which no doubt anyone in time could come to like and zestfully enjoy.

Fromo repeated confidently that his people knew where and how to look and dig for the food roots, how to cure and cook them, and, as he put it, how to take "resurrection" from eating them.

The scholar-chief remarked that while reading history back in college he had gathered that early man lived principally on roots supplemented by fish and from time to time wild game and birds, in great part caught in traps. But roots were the surest and most frequently eaten food. Fromo doubted that man mastered the farmer's trade even to the extent of adapting large-seeded grasses as grain, or of open-herding edible live-stock in very early times.

Archeologists and anthropologists tend to agree that the principal civilizations were "built" on the principal grains, and some contend that they began with roots and worked upward. Fromo made the point, too, that as one with so-called primitive peoples, he doubted that early man lived to any great degree on meat.

His reasoning was that in general and regardless of a hunter's skill, good game meat is extremely difficult to take regularly with bare hands or stone-headed spears or hammers, or with clubs or even primitive traps. Edible roots, on the other hand, were (and are) comparatively easy to find and use, and in many instances they are more nourishing than above-ground harvests or pickings.

There are impressive parallels in present-day agricultural food production. There are now literally hundreds of tried and proved root crops; botanical directories now list almost 700, including plants that bear edible tubers, rhizomes, or bulbs. There are at least sixty other species that are edible by livestock. The listings are probably very far from complete; the agronomy of root crops is still in the borning. Even so, its listings already include several of the most generous of developed food harvests—most generous, that is, in terms of digestible calories producible on one acre or other measure of land surface.

Two of the longest-used staples in terms of feeding people are taro (*Colocasia esculenta*) and cassava (genus *Manihot*).

Taro is a tropical cousin of elephant's ear with highly edible tuberous rootstocks. Despite the usual designations of Tahitian and Maori, the taros flourish in many lands. The same holds true for the even more widely ranging cassava, a great family of tropical plants also with large, edible rootstocks. Measured yields of taro average around twelve million digestible calories per acre (43,560 square feet); cassava yields vary from eight or nine million calories to as many as fourteen million per acre. Either can be relied on to yield four or more times the calories of principal grain crops, and to deliver 75 per cent or more of total plant weights as edible human food—as compared with 25 to 30 per cent for the grains. By averages, granted they are nowhere near high enough to meet nutrient demands placed on them, a "good" grain crop produces from 1.6 to 2.1 million calories per acre in a year. By people-benefiting contrast, the most frequently eaten food vegetable of the Western World, the Andes-born, misnamed Irish potato produces about eight million calories per acre (in the United States, Poland, Holland, and Denmark), and at least six million in most other potato-growing countries. In suitable soils the productive scores for sweet potatoes and yams are close rivals, while the giant-rooted sugar beet is a worthy rival as a famine squelcher, especially since its pulp and leaves make valuable livestock feed.

This quick sampling of food-providing roots or root spheres at least suggests the likelihood that the most and best are yet to come. There is powerful evidence that many of the most nutritious root foods have not even been developed as crops. Dr. Harrison Brown, the eloquent chemist-philospher of Cal Tech, has pointed out that what we most need for feeding the already stifling tidal waves of hungry people is a "good meat beet"—a root crop with a comparatively high protein content. Dr. Robert S. Harris, the renowned biochemist and nutritionist of the Massachusetts Institute of Technology, has spent many years proving that what we also need is better and more exact nutrient analyses, as well as better adaptation to cultivation of the great potential root crops. He submits that competent development of even a select few of these could crush the prevailing thrusts of famine like dropping an egg on the kitchen floor. As one observer of the now momentous progress in plant breeding, I believe that the "meat beet," more literally the higher-protein root, is not only attainable but on its way.

Dr. Harris, as one of the most revered nutritionists of our time, suggests pertinently that now is a good time to bury and forget the frequent misnomer "starch root" as a designation of edible roots as a provident throng. A great many species are not starchy at all, and even of the starch-heavy members, most have numerous and valuable nutrients in addition to starch or carbohydrate, which, heaven knows, has its place in feeding people and stanching hunger.

The readiest example here is the common Irish potato, which is commonly maligned as "fattening." The truth is, of course, that without imposed butter or gravy, a potato is about as fattening as an apple. According to the Department of Nutrition and Dietetics of the University of Wisconsin, this most frequently eaten vegetable has the following nutrients scores: potato pulp contains 78 per cent water, 19 per cent carbohydrate, 2 per cent protein, .01 per cent fat, and .9 per cent ash. On a per-1000-gram basis, it also contains 80 to 110 grams of calcium, 560 of phosphorus, 7 of iron, 13 of carotin, .03 of riboflavin, 1 of thiamine, 12 to 14 of niacin, and 100 to 110 of ascorbic acid. Any way you slice or mash it, or leave it whole, the well-cooked Irish potato is a first-rate, generously yielding food—granting that it is not the most nutritious of known and used root or root-sphere harvests.

American Indians were, and to a measure still are, among the most able way showers or argosy helmsmen in food uses of roots. As a first European colony head of what is now Canada and the United States north of what is now Pennsylvania, Samuel de Champlain, who also doubled as an able botanist and superb recorder of the Northern Indians, noted that even the greatest tribes, including the Algonquins, Iroquois, and Hurons, repeatedly survived people-erasing famines because of their canny acquaintance with edible roots. Even the largest and most provident tribes depended heavily on edible roots.

Of all American sources I know, the most graphically revealing are the *Journals of the Lewis and Clark Expedition, 1804–1806*. Captain Meriwether Lewis, secretary to President Thomas Jefferson, and his sturdy colleague, Second Lieutenant William Clark, recorded in oftentimes sparkling details the momentous dependence of the earlier Americans on roots.

Chosen at random, there is Lewis' account of camping on the banks of the Snake River in what is now Idaho (June 2, 1806):

> After a cold rainy night . . . we sent some men to an Indian village above us, on the opposite side, to purchase some roots. They carried with them for this purpose a small collection of awls, knitting pins, and armbands, with which they obtained several bushels of the roots of cous [cowas] and bread of the same material.

Cowas or cous (*Lomatium geyeri*), now more commonly known as biscuit root or Indian biscuit, is closely related to the cultivated parsnip, and its vegetative kin still grow wild in much of North America and some of Western Europe. A kindred edible root, *Lomatium farinosum,* was included in a recent Massachusetts Institute of Technology listing of the most valuable edible plants of Northern Mexico. This Indian biscuit of the Northwestern United States tastes much like fresh celery and is still eaten extensively by Indians of the Snake River country and elsewhere in the Northwest. It is still used, at least occasionally, for making a highly palatable and nourishing flour. We learn from Lewis and Clark that many tribes preferred the flour for making "travel bread," cakes or loaves large enough to sustain a man during a long journey. There is evidence that in many lands beside the American West, roots have provided the beginnings of bread making.

Lewis and Clark began their food barter with Indians by trading various articles for "yellow loaf bread," made from the root of the wild sweet potato (*Ipomoea pandurata*). This is one of the many truly great food roots of our country; the species grows throughout nearly half of North America, from lower Ontario to Florida and Texas and well into Mexico. One is likely to see the *Ipomoea* growing at country roadsides and in open wastelands or abandoned fields, a modest vine with heart-shaped leaves and flowers like morning glories which are usually white with a pink-and-purple center. The base root of this plant, which is usually found just below the frost line, is quite large; it sometimes grows as much as a yard long and weighs up to twenty pounds. In shape it is like a giant sweet potato, but there is only one root for each vine, and the big lone root lives on year after year, producing new stems each spring. It has brittle, slightly milky flesh, similar to that of cultivated sweet potatoes, but the color is nearer white than yellow; the flavor is less sweet than that of most commercial sweet potatoes or yams. But if it were developed as a crop, its per acre yields would surely be tremendous.

Another Indian root food was the wapatoo (*Sagittaria latifolia*), which commonly grows in wetlands and at the water's edge, particularly in marshes, around the edges of ponds, or in shallow or sluggish streams. The food root is still to be found in much of North America, and kindred species are listed in much of Europe and Central Asia. The plant called tule is now on the official crop list of the Republic of China (as a duly approved field or garden crop). This family or genus is one of look-alikes, arrow-shaped leaves, white flowers about an inch wide and usually found in clusters of three, and smallish edible tubers frequently no bigger than a hen's egg. There is considerable similarity to the common potato, in both taste and appearance, but many people find the wapatoo the better flavored.

Meriwether Lewis wrote that he and his party survived the long hard winter of 1805–06 mostly on a wapatoo diet.

> They are never out of season. . . . They are nearly equal in flavor to the Irish potato and afford a very good substitute for bread.

Both by direct statement and by lucid implication, the eminent chronicler of the Indian's West indicated his belief that without the many kinds and sizes of nourishing roots available, few if any of the earlier Western Americans could have endured. Their fishing was sporadic, game was sometimes scarce, and many of the Indians were appallingly bad hunters. In instance after instance, edible roots were their staff and salvation.

Interestingly, many contemporary American Indians, both on and off the Western reservations, still rely on native roots to supplement and improve their diets. In the main correctly, they take for granted the nutritional merits of the roots; their gourmet's gift lies largely in knowing which ones are best eaten raw and best suited for grinding into flour or for sun drying or cooking *au naturel*. Interestingly, too, many of the edible roots and at least some of the cookery techniques described by Lewis and Clark are still used by our Western Indians, certainly including the Sioux and Navajos.

One noble example is the still-wild-growing root that Meriwether Lewis called Jerusalem artichoke. One of my botanist friends points out a shade waspishly that this is a double misnomer, since the plant is really a sunflower (*Helianthus tuberosus*) and has nothing to do with Jerusalem—which is a malapropism for *girasole*, "turning with the sun." In any case, the food source is a handsome perennial with hairy leaves and typical sunflowers about three inches in diameter. Its multiple creeping roots bear long, flat, highly nutritious tubers of excellent flavor, still very useful as a soup vegetable or a food for invalids, as well as a staple entree. This *Helianthus* flourishes in lands too dry and hot for the white potato; it has an impressive record as a long-time garden crop of the North American Indians, and it could very well be developed as a superior dryland crop for much of the hungry world.

Lewis also studied another edible root which was eventually named (in his honor) *Lewisia rediviva*. This plant graces many of the foothills and shelflands of the Rocky Mountains. Its unflattering popular name is bitterroot, but anyone who samples its richly flavored white taproot soon finds that the bitter taste is only in the peel. The *Lewisia* is a vigorous perennial, and this fact, together with its nutritional worth, adds to its status as a lifesaver for mountain travelers in need of food. It is stemless, with narrow oblong leaves which grow directly out of the top of the fleshy, carrot-shaped root. White or pinkish blossoms, arranged in a wheel, rise directly above the tuft of leaves and like those of *Helianthus*, turn with the sun. *Lewisia* is now the state flower of Montana.

Meriwether Lewis also heeded with deepest respect the prairie apple or breadroot (*Psoralea esculenta*), which remains one of the great traditional foods of the Sioux and other Great Plains Indians; it grows from Manitoba to Texas. The root is starchy and glutinous and has such an excellent flavor that it is palatable raw as well as baked. Breadroot belongs to the pea family and it can be recognized by its purplish-blue flower, resembling a sweet pea, on an erect, slightly branched stalk. The root is perennial, but the top usually dies off in late summer so that the best harvesttime is midsummer. The root yield is generous, and there is evidence that the plant has been a cultivated crop. The traditional Sioux method of preparing the roots (still practiced to a degree) is to braid them, dry them in the sun, then grind them into a meal which makes a superb dressing for buffalo, venison, or beef, or for mixing in succotashes. Here again, there is reason to believe that this plant would be very valuable as a cultivated food crop for low-rainfall areas.

Ipo or squawroot, sometimes known as yampa (*Carum gaird-*

neri), is still another valuable native food root that seems entirely worthy of being established as a standard crop. A perennial of the caraway group, it is indigenous to most of Western North America from British Columbia to lower California, and eastward as far as the Black Hills of South Dakota. The fleshy, carrot-shaped roots usually grow singly, but sometimes in groups, particularly in fertile soils. They have a sweet nutty flavor and are excellent with poultry or fish. Squawroot is quite delicious as an entree, too, especially if it is slow baked and served with plenty of butter. Like other great food roots, squawroot makes excellent bread. After collecting the roots, preferably in late summer, many Indians and others put them to soak in water, remove the brown bark, and dry them for grinding or pounding into meal or flour. From either form comes delicious high-caloried bread or cakes.

One of the first American food crops mentioned in the English language was the Virginia tuckahoe (*Peltandra virginica*), which was noted by Captain John Smith of the Jamestown Colony as one of the arums. The tuberous food plant grows in wetlands, usually at the water's edge, all the way from Eastern Canada to Florida. A similar species flourishes in comparable areas throughout Northern Asia and much of Northern Europe. The roots frequently weigh four pounds or more and are perennial. John Smith recorded that the Virginia Indians roasted them in pits, then dried and pounded them into a palatable meal. In Northwestern Canada the tuckahoe is still used for making bread. The Seminoles of the Florida Everglades still harvest coontie— the large, starchy root of the native cycad (*Zamia floridana*) —and use it to make excellent bread.

One who would hunt down edible roots, even if only as a hobby or outdoor sport, soon finds that in instance after instance there is no fixed growth range; that identical or closely similar species can be found in any one of dozens of different countries —sometimes on different continents. The student of edible roots presently finds himself hip deep in herbaceous internationalism, which seems to say: "Look below you to feed the hungry—not in just one community but in a hundred nations."

A great many records of American settlements in early colonial times tell how white people and Indians alike frequently depended on the roots of the still common wetland cattails (*Typha latifolia* and *Typha angustifolia*) for defense against deadly hunger. It is easy to find out that these roots are very presentable food. The surprise factor is that the bearing plant, which so many generations of Americans have termed native, is now listed as native to Central Asia. Somehow, in the wonderful ways of vegetative migrations, it reached North America well ahead of the great rush of European settlers and gained natural establishment in marshes or other wetlands throughout at least half of North America.

Cattails are natives of at least two continents; other edible roots can be found thriving in even greater portions of the earth. Among the better known and more nearly global examples is the so-called evening primrose (*Oenothera biennis*). Its sizable perennial roots provide a partial substitute for potatoes; the young green shoots make excellent salad greens. Champlain mentioned that this root crop was first brought to North America, in the early years of New France, from Old France, where the plant had been a cultivated field or garden crop as long as anyone could remember. In Germany today the plant is known and grown as German rampion. Yet the same species is findable as a wild-growing resource from Labrador to Florida, and at least as far west as the lower Mississippi Valley.

Again, sad to relate, this very noteworthy potential food crop is not being developed as a food crop even at this time when the world at large is being confronted by momentous tidal waves of hunger.

A worthy companion is the so-called sweet flag or calamus (*Acorus calamus*), also one of the most widespread of the edible root crops. It continues growing wild throughout most of North America and Europe and in much of Asia. The rootstock or rhizome is a superbly flavored, nutritious roasting vegetable with a host of incidental food uses. There are many other clans and species of edible roots and near roots still to be found on a hemispheric basis. Among others are the wild licorices (*Glycyrrhiza*), long-lived perennials with long, fleshy roots which are richly flavored and highly nutritious.

Hardly less international and intercontinental are the wild-growing edible bulbs, which offer substantially the same story, but, alas, from another book. Among the probable hundreds of species at least a few have found places in decisive American history. You may recall how Père Marquette wrote that wild onions, which he learned to like while visiting Indian camps and villages, saved him from starvation. Or how the providently edible corms of the sego lily (*Calochortus nuttallii*) held back starvation from Mormon pioneers on their arrival in Deseret. (It is understandable that the sego-lily blossom is now the official flower of Utah.) There are the highly edible bulbs of the wild yellow lily, the aloe, the star-of-Bethlehem, and on and on through chapter, book, and bookshelf.

The real point to all this, and it may be a crucial and people-saving point, is that through the centuries edible roots have fed the hungry and saved the starving when other sources failed. There is no real cause for doubting that they can do it again. There is very special comfort and significance in the fact that big roots, many of them tuberous, markedly help the very special talents of their plants to adapt to an immense variety of en-vironments, by providing exceptionally generous and living res-ervoirs of foods.

Not long ago Davis W. Cloward, one of our new generation of highly practical agronomists and plant explorers, had op-portunity to reflect on this very important fact of life and nutri-tion. He was plant hunting in New Caledonia for the U. S. Department of Agriculture. His immediate mission was to locate ancestral species of bananas which still have viable seed—essential to some important hybridizing experiments. In a remote, fogbound valley of New Caledonia, Cloward met and made friends with what he describes as the best-fed and healthiest-looking native community he has ever encountered in all his South Pacific travels. But where did the good food come from? The plant hunter could see no livestock in the area. He had earlier noted that fishing and game resources seemed poor. The only sizable crop cultivations he could see were some patches of rather scrubby-looking bananas.

Bananas, as United Fruit and other well-heeled prophets have permitted it to be known, are tolerable food. In Uganda, Tanganyika, and other tropical African places (surprisingly enough, at least half of the world's banana supply is grown in Africa) one meets agrarian tribes that live principally on the ancient crop. But Cloward could not believe that the planting he viewed would yield enough even to begin to feed the community.

In highly developed banana-growing areas of Central America and Ecuador, favored commercial varieties of bananas, such as the Valery, now yielding between 600 and 800 bunches per acre yearly, very probably supply more digestible calories per acre than any other major above-ground crop: some nine million calories per acre yearly. However, subsisting almost entirely on bananas requires some five pounds of the fruit daily per person, and this onetime banana chaser would include this in the God Forbid or Perish the Thought Department.

Davis Cloward shortly discovered that the bonanza of good food was coming from *below* the surface, not from above. The less than impressive banana plantings he viewed were those of the so-called giant root (*Musa leanthes*), whose underground rhizomes handsomely justify the name. Shaped like colossal turnips, the bulbous roots each weigh from fifty to eighty pounds or more. And they, not the mediocre fruit, are the prime food providers. When dug, washed, peeled, and roasted or baked, the gold-yellow flesh tastes considerably like sweet potatoes but, as Dr. Cloward testifies, considerably better. Preliminary nutrient tests show food values definitely superior to those of banana fruit (granting the nutritional value of the latter is quite commendable and quite similar to the Irish potato). But the giant root seems to have a substantially better food value, less sugar, and more food minerals, niacin, and aminos—the protein bases. The calorie production, which is not finally verified (even though *Musa leanthes* is a widespread subspecies), is stupendous: quite possibly above twenty-five million calories per acre, more than a dozen times the average of our principal cultivated field crops, and three times the average of superior potato yields.

The Caledonian growers, by the way, ably demonstrate a simple and effective technique for perpetuating the bounties of underground providence. When they dig up the "mighty turnips" they invariably cut out a "bit" of the rhizome, with a living "eye," or bud, and a stalk of the shoot, or true stem, and replant it for future growth and use.

Beginning in the late 1940s, Dr. Robert S. Harris, the foodplant exploring director of the Laboratory of Biochemistry and Nutrition at Massachusetts Institute of Technology, with the help of a superbly competent field staff, spent several years discovering and making exact nutritional analyses of edible plants in Mexico and Central America. The work turned out to be one of the strongest and best-detailed tributes ever paid neglected food roots. In all, the Harris research group discovered eighty-six potential edible root crops in the one subcontinent, which is only about one fourteenth as big as the United States, not including Alaska.

Among these are several particularly significant multiple-food providers. Included is the perennial and readily findable chayote —more formally, *Sechium edule*. Though not as yet a major cultivated crop, the chayote can frequently be seen in yardways, garden edges, field borders, and fence rows.

It may well be the most completely edible of known food plants. The pod or fruit is tasty and exceptionally nutritious as a cooked vegetable. The young leaves make an excellent salad green. The flower is also edible; it provides an excellent flavor entry for egg dishes and salads. But the most nutritious of all the edible portions are the roots. Peeled of their handsome brown skins, they are a crisp, succulent, splendidly flavored, and, as the M.I.T. survey shows, splendidly nutritious food. The reminder here is that quite possibly those plants that are edible from the roots up are the very special hopes for man's survival during the leaner decades ahead.

The late Dr. H. W. Staten, whom we have also mentioned earlier as a great teacher of agronomy and who ended his career as agricultural advisor to Ethiopia, ventured into the Upsante country in the rocky, dry foothills of southern Ethiopia. There, like Davis Cloward in New Caledonia, he was surprised and pleased by the healthy appearance of the people. An elder told him that the exceptionally good health of his tribe was due to their having plenty of a particularly good combination of foods—goat's milk, goat's meat, and "fine-o" green cabbage.

Although he could see and smell the plentitude of goats, Dr. Staten could not believe that cabbage, certainly the kind he knew, could thrive in the dry, rough soil, fertile as it no doubt was.

He was presently shown the native cabbage source. It is a rather odd-looking, stubby shrub, a long-lived perennial, which bears a generous number of compact leaf heads. These are generally similar to our garden cabbages, favorably crisp, and of excellent cooking and eating qualities. Though smaller than ground or row cabbage, the bush cabbage bears far more generously and yields edible leaves throughout most of the year.

The wonder factor, as Dr. Staten recorded it, is the very hardy roots, which apparently in the course of prolonged struggles to penetrate the hard, dry earth have somehow developed or "mutated" from a biennial plant (like our common garden cabbage) to a long-lived and far more prolific perennial. The Upsante cabbage bush, now duly recorded in British botanical directories, is still a cabbage, but by way of the timeless magic of roots it is now a transformed cabbage.

There are many other such plants. One was recently noted by the Harris-M.I.T. survey of edible plants of Middle America. This is a highly edible species of the chile dolce, one of the very much worthwhile (and too often overlooked) capsicums or sweet peppers, all of which are believed to be native to lower Mexico or upper Central America. Rather recently this wild-growing dolce, apparently by means of self-generated root hardiness, has changed itself from an annual to a perennial. The fact that it bears edible pods or fruit throughout an average of six months a year and is adaptable to a versatile climatic range has prompted agronomists with the Guatemalan government and the national agricultural school to instate the chile dolce as a particularly promising food crop worthy of distribution to citizen farmers.

The Staten hypothesis is that with improvement of root strength, some, perhaps most, of our annual plants can be "converted" to perennials which in due course can and will bear a great deal more food more dependably than most of our annual crops.

Perennial plants comprise the overwhelming preponderance of all the now known vegetable kingdom. Many plant historians believe that the annual crop is mostly a man-devised or -directed phenomenon suited to man's migratory habits and expediences. Many of our annual crops are believed to have had perennial ancestors; the tomato, to cite just one example, is descended from a perennial herb native to South America.

The advantages of perennial food plants over annual crops are quite evident. They include the savings in labor, seeds, and other planting stock requisite for annual or seasonal planting; the savings in erosion losses incident to reopening the soil every year or growing season; and the formidable waste of fertility and growing time incident to waiting each year while seedlings grow to size and strength for bearing harvests.

The foregoing, obviously, is not directly related to edible roots as such, but rather to the definitive role of roots in producing food. But the interrelations here are openly challenging, and as Dr. Staten so ably noted, they join in properly emphasizing the ever more absolute mandate that man's chances for survival lie with increasing certainty in the provident realm of vegetative roots. For these underground miracle makers not only supply people-sustaining foods, they perpetuate and multiply the staff of human and most other animal life. The already proved capacity of roots to live on through years, decades, and centuries provides a basis of fact for such snatches of future conversation as: "Show me the way to the pea vineyard and the tomato grove," or, "I'm only planting some flowers for the wedding of a granddaughter who hasn't yet been born."

For the world at large the momentous importance of edible roots and the life-sustaining necessity of more vigorous crop roots cannot be overstated. The shadows of hunger now darken the greater part of our troubled earth. The highly respected Food and Agriculture Organization of the United Nations estimates that most of mankind, some two billion of the 3.6 billion people now living, are less than adequately fed. World population currently increases at about 2 per cent a year; world food production increases by a hard-strained average of barely 1 per cent. What Mahatma Gandhi used to term the unending involuntary fast continues to invade the continents and populous islands with the highest prevailing birth rates—Africa, Asia, South America, and, even closer, the subcontinent of Central America and the people-crowded Caribbean islands. The onslaught of hunger is now reported in more than a hundred nations, including most of the sixty new and developing nations which have come into being since the close of World War II.

With terrifying forthrightness, the Food and Agriculture Organization confirms that about 70 per cent of the world's population under six years of age (in all, at least 350 million babies

and young children) must now be listed as undernourished—
a condition ranging from a chronic shortage of necessary nutrients to aching, killing hunger. Meanwhile, by global averages,
the population explosion continues; world census, which has at
least doubled since 1900, will inevitably double again by A.D.
2000.

The stork continues to win over the plow. If this goes on,
mass starvation, wars, rebellions, and interminable civil strife
wait as lethal harvests. The prevailing contest between people
and hunger, which people are presently losing, is everybody's
contest. None can evade or run away from it. For the greater
part of the world, food demands are already playing havoc
with supplies. There are no literal surpluses of food anywhere;
and in a countable majority, people—the most numerous of the
higher animal species, now far more numerous than cattle,
swine, sheep, or other comparable animal lives from aardvarks
to zebras—are living below the nutritional danger lines.

With five eighths of mankind now lacking minimal food supplies, a fourth no better than self-sufficient in terms of food
production, and only one eighth with food to spare, the need
for doubling, tripling, quadrupling, and quintupling food production is now absolute and inescapable. Roots, more and better-
propagated edible root crops, and more vigorous and provident
root sustainers of above-ground harvest are the great, the mystic,
the ever-burgeoning hope and promise of solution.

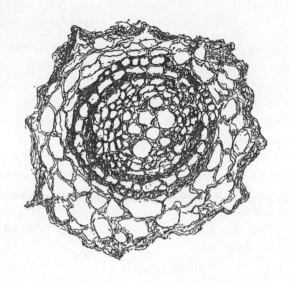

A CLOSER LOOK

Here on my workbench is a well-rooted strawberry plant which I dug earlier today from the side-yard garden of one of my more forgiving and later-sleeping neighbors. I chose the strawberry because its growth range is exceptionally global and because its root structure is reasonably typical of many other exceptionally interesting and valuable perennials.

I have borrowed this particular strawberry plant for still another reason. It—better say, she—is a propagated descendant of a herbaceous multigreat-grandmother, with a root system of most impressive vigor. This wonder thing came into being—just "spanged out," as the local saying goes—on a remote Vermont hillside. This remarkable sport, or self-arrived mutant, so markedly different from its parents and kin, was presently noted by an exceptionally sharp-eyed stroller. At the time (on an early autumn day in 1938) George David Aiken, now the senior Republican member of the United States Senate, was the hard-working young Governor of Vermont from Monday morning through Saturday noon. He was spending the rest of his time at home on his Putney hillside attending the homestead nursery which he and his bride had founded in 1919.

On that particular Indian summer day in 1938, while exploring his back pastures for native wreath materials, the gubernatorial nurseryman spied a wild-growing strawberry plant with a very large and flashing red berry. Having happily eaten the surprising tidbit, the Governor used his strong bare hands to scoop out the plant which he painstakingly reset in a deep-soiled edge of his wildflower garden.

After weeding, tilling, subdividing, and otherwise coddling the transplant for nine successive years and assembling its progeny in a single bed, the then Senator discreetly announced (in his annual nursery catalogue) the official advent of the Aiken Ever-bearing Strawberry. There have been "everbearing" strawberries since goodness knows when; the designation merely indicates a variety that produces a late or bonus crop following the usual main crop. There are some hard-to-explain genetics involved, but the several-times-great-grandmother strawberry has the momentous advantages implicit in an exceptionally large and obviously hardy root system.

The Aiken Everbearing is further notable because it presently became the first variety of a commercial crop to be granted a real, honest-to-goodness U.S. patent. But its more definitive advantage was and is exceptional root hardiness.

This, plus the likelihood that I could take from the very loose soil most of the root structure—without being caught in the nefarious act, had prompted my early-morning borrowing. To spill the whole truth, there was also a factor of reciprocity involved. In late winter, the now elder statesman of Vermont had been snitching, grafting, and budding "wood" from my

ancient sour-apple trees. And science, so it says here, is a phenomenon of take and get taken.

In spite of my devious planning, the strawberry root structure
here on the table is probably nowhere near complete; lifting
the entire underground portion of a sizable perennial is next to
impossible. Strawberry roots are built to last; quite conceivably,
this particular system could live at least as long as any man or
woman. The "long" has double meaning.

The root system here before me has a down reach of about
twenty-one inches, a side reach of about nineteen inches, and,
counting both main and feeder roots, a combined length of at
least 130 feet. The above-ground portion is a shade more than
four inches high; the length preponderance of the roots shows
for itself. I reflect again that generous harvests and generously-
sized root structures usually go together; folklore to the contrary, healthy, ample-sized roots in fertile, well-textured soils
do not necessarily "go to tops."

The strawberry plant here on my workbench exhibits other
distinctive resources of valid roots. One is that a root system
usually if not invariably has its own distinctive scent. Strawberry roots have a slightly pungent smell which suggests a
faintly perfumed toilet soap. Like most other vegetation, the
strawberry plant stakes its claim to its own soil area with the
distinguishing scent of its roots. There is evidence that many
of the undersurface enemies of roots, including the destructive
threadworms now called nematodes (which happens to be
Greek for "threadworms"), locate their quarries at least in some
part by the telltale scents of roots that are especially tasty to
nematodes.

Quite evidently the scents that roots leave in soil come from
complex chemical compounds which the roots themselves manufacture and exude. These exudates, which apparently include
growth regulators, sometimes move from the roots of one plant
into those of adjacent plants and influence them sometimes for
their benefit, sometimes for their hurt. One of the more defini-

tive studies of root odors was instituted back in 1954 by Dr.
P. J. Linder and Dr. J. W. Mitchell of the U. S. Department
of Agriculture Research Center in Beltsville, Maryland. The
Linder-Mitchell finding is that most root structures, like many
animals, use distinguishing and self-exuded scents as a means
for staking claims to living ranges or "homesteads." Probably
this strawberry plant I have borrowed has other means, no less
subtle and ingenious, for staking its special claim to its homeland; but its subdued odor is quite distinguishable.

I get out my microscope for a closer view of this fascinating
understructure which is a strawberry root. With a razor blade
I cut a paper-thin cross section from a feeder root and place
it on a slide beneath the lens. The microscope is an old-timer
the worse for a great deal of rough trouping. Even so, a glance
shows clearly the consistent structural pattern in this thinly
sliced cross sector of a strawberry root.

The primary pattern is established by living cells held in
place mainly by a mortar of pectic materials with some lignin
present. Looking more carefully, I see at the center a circular
area containing a number of patches joined together in a design
that suggests a star. This is the xylem (from *xylon,* Greek for
"wood"). The cells of the xylem form long up-and-down tubes
which apparently fulfill the function of carrying upward the
various fluids, including water and mineral solutions. Since these
feed tubes are also stiff-walled and extremely tough, they add
strength to the root structure as a whole. As I view them in
the cross section, the living feed tubes look more or less circular.

In the center part of the root between the patches, or points,
of xylem are groups of very small cells, which are also tubes
but thin walled and filled with protoplasm. These compose the
phloem, the function of which is to conduct carbohydrates
downward from the leaves. Thus, quite clearly, roots live the
rules of one-way traffic; the upbound and downbound traffic
are kept apart.

Barely visible even with a fairly competent microscope is a

line of cells between the xylem and the phloem. This is the cambium—latin for "change." Here the cells are constantly splitting in two to produce new growth; the new cells are divided between the conducting tissues on either side. Between this central area and the epidermis, or outer skin, of the root is a mass of cellular tissue, looking in cross section like roughly laid tiles. Here and there between the tiles are air spaces. Water absorbed from the soil moves through these cells to the inner parts of the root, and food is stored in this tissue.

I again reflect that among the very special masterpieces of the living root is the root hair. Just behind the tough, burrowing tip of the root and each of its branches is a small region covered with hairlike outgrowths, each from a single epidermal cell. These absorb almost all of the water that is taken in by the roots, and they are continually disappearing and being replaced as the root pushes farther into the soil. Some root hairs live for only a few hours; in the case of our strawberry root, as nearly as I can estimate, few if any live for more than three or four days. To the amateur with a microscope, looking for root hairs is like an unending, sometimes exasperating shell game: there they are and there they aren't.

I clip off one of the tinier root tips and place it under the microscope. Right behind the drilling tip I see the newly formed root hairs, apparently several hundred to a millimeter or, say, one twenty-fifth of a square inch of surface. For the most part they appear to be engaged in very purposeful work. Many are wrapped determinedly about particles of humus. Clearly, they are serving as anchors; some additional microscoping shows they are also absorbing moisture which probably contains dissolved nutrients.

There is no way to separate literally the function and the structure of roots—as the textbooks say, their physiology and histology. The two cling together like molecules of water Englishmen mislaid in Ireland, or Arkansawyers dropped into New York City. But unlike the clingers together just mentioned, these

roots must keep endlessly at growing, pushing into new areas
and depths of soil or, at the very least, permeating more completely their available soil spheres. To accomplish this, the cells
continue to divide, and thus produce still more rootlets and root
hairs.

As this strawberry root system proves, this extension is especially visible in the feeder rootlets. Hundreds are materializing in the section I am now viewing under the microscope.

In the late 1920s one of the great authorities on the diseases
of sugar cane, Dr. Raphael Ciferri, won notice and in time
followers by his assertion that aggressive root growth is the
prime requirement of healthy vegetation, and that any factor
that hinders the rapid formation and growth of new rootlets
invites an outbreak of ruinous vegetative diseases or a destructive invasion by other soil lives. His view was promptly and
aggressively attacked, but subsequent research has supported it.
The Italian-born pathologist was an especially able student of
harmful root fungi. One of his special areas of research was
the fungus genus *Phytophthora*, which accounts for the "wet-
feet disease" of strawberries, the "under-rot" of pineapples, the
"decline" of avocados, and other so-called wet-root diseases.
Rather typically, the enemy fungi concentrate on new rootlets
that have been weakened by tenancy in excessively wet soils.
The root system itself sometimes attempts to combat its unfavorable condition by producing anaerobic rootlets—rootlets
that can manage without the oxygen contained in drier soils.

As far as I can see, my borrowed strawberry plant has not
been obliged to perform this last-named sorcery for defense of
its roots, but in any case, this ability of root systems to produce
special types of rootlets promptly in serious emergencies is still
another wonder item in our story. Behind it, perhaps, lies the
solution of the ancient mystery—the manner in which plants
managed to take leave of the sea, where they are believed to
have lived for hundreds of millions of years, to grow on land.
A paleobotanist has suggested that these anaerobic or oxygen-

independent rootlets date back to the time when all life took its oxygen from water and all earthly life was essentially marine life.

Why very wet soils are so harmful to roots is a baffling question, and one of great consequence. Long ago (beginning as far back as 1914) J. W. Shriner and A. Skinner, of the U. S. Department of Agriculture, offered solid proof that arrested or otherwise inferior root growth is related not so much to excessive water as to the accumulation of chemical sludges in wet soils. The findings of these men were later supported by an investigation in England of the so-called midsummer slump in yields of greenhouse tomatoes.

One of the outstanding American students of root life, Dr. Stephen Wilhelm of the University of California (Berkeley) faculty, believes that the living sphere of the root decides the fate of vegetation, granting that within the root area, or rhizosphere, life and death are continuously and permanently interrelated.

Plant roots provide the tenancy for numerous alien organisms which cause them to sicken or die [Dr. Wilhelm explained to me in an interview]. Root secretions, carbon dioxide, rejected transient root parts, such as root hairs and root-cap cells, figure strongly in the interrelated life spheres of the root region. . . . The prospering of root structures has both direct and indirect relation to neighboring soil lives. A typical saprophytic fungus (one which lives on dead organic matter), while growing on the root surface or in its rhizosphere, may obtain all or part of its nutrients from the secretions of the living root. Although the fungus' growth may depend on these, the fungus itself is actually an outsider, supported, not from the table of the host, but from the garbage pail. Even so, the fungus may injure the rootlet host by its presence, even though it does not actually feed on it.

The living root system is host to what may be called a select biological community of soil microorganisms that also feed on the figurative garbage of the root system. But the fact remains that, whether outside the rootlet or inside, waste-eating fungi and other soil organisms can cause root diseases, though many do not. The root structure is also under never-ceasing attack by directly parasitic fungi and other disease bringers that penetrate and infect the living tissues of the root. The structure of roots must, of course, bear with the fact that root life is everlastingly in line of direct attack or indirect injury by fellow residents of the soil. . . . It follows that the root system is forever on the defensive, ever destined to be the most disease-prone part of vegetation, and sickly as the animal kingdom sometimes appears, it's always a safe bet that the vegetable kingdom is sicker. On that basis, it cannot endure without growing, cannot hold its own without continuously moving ahead.

The Wilhelm view, which is supported by many other scholars, stresses that the continuous dying of cells and of specific parts of a root structure also opens the way to infection. Fungi are especially disposed to seek escape from the fiercely competitive soil environment by taking refuge in the weak areas of roots. It follows that plant enthusiasts are perhaps too inclined to appraise root health by the appearance of the leaves and stems of the plants they support. Some hold that this is justified because the functions, growth, and general health of roots are largely dependent upon carbohydrates prepared in the leaves. However, the more recent consensus of other plant physiologists and pathologists is that the actual extent of dependence of roots on leaves—granting that there is dependence—has been greatly exaggerated.

It is the seeds and shoots, not the roots, that have first claim on the important products of the leaf. At best, the root system is third in the receiving line for the sunshine bounties. Even though vegetative life and growth involve a mutually beneficial barter between the roots and the above-ground portion, there is incontrovertible evidence that roots almost invariably give more than they get. Certainly in the case of perennial plants—and it has been estimated that certainly more than 99 per cent of known plants are perennial—the leaves are shed periodically and the above-ground parts take a rest. But the roots of perennials have the capacity to work around the clock and the calendar. When winter or (in the tropics) leaf-shed dormancy ends, roots must resurrect the plant, energize the buds, and sustain the growing leaves until they are large enough to carry on photosynthesis. Obviously, the work load varies with the season, but if the work of roots ever stops, the plant inevitably dies.

The examination of my strawberry root makes me wonder how much of the work it does is purposeful and how much is the result of sheer happenchance. Some plant students insist that roots do not actually grow (or go) after water and other plant nutrients but merely stumble upon them. Several of my naturalist friends agree with this view, but my plodding layman's experience makes me think otherwise.

The antics and revelations of my shoddy little microscope remind me of a vastly more competent lens which enabled an inquisitive Italian to take motion pictures of another system of strawberry roots. The film, apparently made by a member of the Italian army signal corps, came to my desk as part of a cache found by American forces during the Salerno campaign of World War II. Much of it had been damaged but what remained was a little masterpiece. The artist-photographer signed his work "P. de Marco."

The undamaged part of the film had no sound track or printed commentary, but consisted of a succession of stop shots made from an excavation of soil and subsoil. The record began in the springtime and included summer, fall, and winter sequences, and then part of the following spring.

The first sequence gave credence to the folkish belief that roots do indeed have eyes. It showed roots growing as though they knew their direction, circumventing all manner of obstacles, and appearing to "see" around corners. Confronted with a peb-

ble or small rock, the growing rootlets were shown skirting
deftly around the impediment. When a larger obstacle such as
a fieldstone, the butt of a fence post, or a mass of competing
roots get in the way, the root growth slowed down before an
impact occurred. In one instance, when there was no way
around the obstacle, the brush of rootlets simply stopped growing. But usually the roots dealt with obstacles by changing or
even reversing their direction or by zigzagging around them.
Confronted with deposits of sand or sterile clay, a few of the
rootlets pushed through dutifully but others turned toward
more favorable areas.

Part of the initial sequence of the film was centered on rootlets that were developing into feeder roots. A close-up showed
the primary structure as a slender brush about two inches long.
Its threadlike side branches first appeared to grow into position
as anchors. The early growth was seen to be quite vacillatory;
repeatedly the extremely fine rootlets disappeared and were
quickly replaced by new ones. The more permanent root structure that presently emerged grew almost straight down. The
revealing camera followed the longest of the down-going roots,
showed it making deft turns to by-pass pebbles, and from time
to time shifting its course as if to take in some tempting bits
of humus.

Above the drill cap of the prime root, one could see tiny
root hairs which occasionally appeared to move from side to
side like small untended hoses. Their color varied; most were
tan or light brown, though a few were gray, lavender, and even
green. As the taproot kept drilling deeper, it produced branches
or clusters of feeder rootlets no thicker than very fine thread.
Some grew rapidly, others slowly, and still others were dying
and sloughing off.

A panorama sequence indicated a dry period. This served to
stimulate root growth. The drilling tips seemed to harden and
they burrowed downward more and more rapidly. As the dryness continued, the growth of lateral roots near the surface

seemed to slacken. This seemed to spur the deeper roots to even greater activity. They encountered hardening subsoil, but their pinkish drilling caps still pushed on. Perhaps because of the summer drought, the length of the central root increased almost a foot.

Rain came. The entire root system, particularly the parts near the surface, resumed rapid growth. Several ominous fungus shapes appeared. Lateral roots were constantly dying, but usually were promptly replaced, and the secondary roots kept growing in all directions at every imaginable angle. The leaves of the plant resumed their growth, but as autumn moved in, practically all activity was underground. Leaf growth stopped quite abruptly as winter came, but the roots continued to grow, though not so rapidly as in spring and summer. In the late winter there was a rather sudden increase in the number of root hairs.

Then came the rush of spring. Uptilted panorama shots showed an impressive growth of leaves and surface runners, and then the powerful lens returned to the roots to show them pushing on vigorously, growing even faster than the stems. The sequence ended rather abruptly with a great lateral sweep of the expanding roots and a flutter of damaged film.

The pictures only intimated, and that quite incidentally, the foul play of pathogens. However, for brief sequences there was evidence of a retardation of growth in particular areas and of casualties among the feeder rootlets. But along with this was shown the prompt replacement of the lost roots. The film suggested that danger to the plant was centered in the feeder roots. Where subjected to mishap, such as flooding, they died very quickly.

Later a nurseryman, whom I know and learn from, substantiated the hint about the danger of flooding roots. When watered too freely, he said, small roots of snapdragons, for instance, die in less than an hour. He also pointed out that injury from overwatering tends to be most severe in soils with a high proportion of organic matter. And here he noted that most plants have an almost miraculous talent for meeting wet conditions by forming new rootlets which are resistant to excessive watering. Emergency mud roots are usually shorter, less branched, and softer skinned, and have less insulation than normal roots. After a thoughtful pause, the nurseryman added that roots, at least many of them, are amazingly adaptable to given soil environments, and that some adapt themselves to situations wholly removed from the soil—tropical orchids, for instance, which regularly grow in the air and absorb rain as it falls.

Until comparatively recent times roots were taken more or less for granted. Their main function was believed to be to anchor the plant above, hold it upright, and somehow supply it with water from the ground. We know now that roots gather not only water but mineral nutrients from the earth and that they perform astounding chemical feats.

The roots of perennial plants and of certain hardy annuals live through the bitter cold of winter even if they freeze. The winter wheat that shows up green through melting snow has roots that have sustained it through the long cold months and will enable the grain to grow and produce an early crop. Were it not for root survival over winter, strawberries would have to be planted every year. And how bare our flower gardens would look in early spring if it were not for the roots of the perennials. For who could work the frozen topsoil to plant crocuses and snowdrops in time for blossoms in March and April? Pondering these matters, I return my strawberry plant, with deference and esteem, to its home in the fringe of the Aiken wildflower planting. What the Senator doesn't know doesn't hurt him, and the senior Republican from Vermont is not often hurt.

Most roots begin from seeds, and we can observe this beginning quite easily. For one engagingly trivial example, I have on my desk a newly sprouted oat grain. The seed is well developed and big enough to show clearly the first little struggling root—the beginning of a complicated underground system which will anchor and nourish the young plant.

Last Friday I left my oat seed on a scrap of wet blotting paper. Now it is Tuesday morning and, rather miraculously, the oat seed has become a living plant. Its tiny root has responded to a remarkable urge (somewhat frustrated in its present position) to grow downward, and, at the opposite end, the emerging shoot shows an equally implicit urge to grow upward. It is easy to see how the root and the shoot are joined together in the seed which has sheltered them. The endosperm, packed with nutrients for both the oncoming root structure and the top, is the most conspicuous part of the seed. For the time being, the food reservoir is becoming even larger as it absorbs moisture.

A quick look through the magnifying glass reveals that the shoot portion is a tiny bundle containing neatly rolled leaves. A hillside neighbor, who used to teach chemistry at Harvard, has told me that a vital chemical compound is activating the leaves, a growth-influencing substance or hormone called indoleacetic acid. How does the hormone get there? The answer seems to be that the parent plant makes it and puts it there. It is probable that the plant manufactures the compound from a protein-related acid called tryptophan, and places the finished product in the seed.

All these near certainties support the thesis that plant life begins, ends, and begins again as an extremely efficient chemistry laboratory. Roots are not the only part of the plant that is in on the act, but they play a leading and indispensable role. Chemistry takes over the plant's development the instant germination begins. The enzymes, the organic catalysts inside the seed, seem to leap to the task of breaking down the stored proteins.

As germination begins, growth of the embryo plant is accomplished by division of the cells. The hormone seems to prepare the way by softening the cell walls so that the cells can

absorb water. The nucleus of a cell divides, and a wall is raised
between the two parts by means of the pectin contained in the
cell. The pectin fluid and perhaps other natural plastics, plus
cellulose fibers which now materialize, give rigidity to the cell
walls, though they can be softened again by action of the plant-
made hormone. This, as the manuals reassure me, is only one
of the many remarkable functions of indoleacetic acid. Another
relates to directing the root portion of the plantlet to grow
downward and the shoot upward.

Now I plant my engaging but baffling oat seedling on the
fringe of my vegetable garden where it will probably continue
to pose profound questions. At least, another young plant life
is on its way, raising its brave new leaves, while below the
surface those mute alchemists, the roots, are performing their
brilliant and, as it seems to me, magic chores.

Even so casual a glimpse of an oat seed producing a living
plant suggests that from the moment of a plant's birth the root
may be more intriguing than the stem. Yet it cannot be separated
functionally from the rest of the plant.

One also observes that water is indispensable to all parts of
the plant from the time of seed germination. Every cell is
given a share so that its components can function properly.
Part of the water that the roots absorb is used in their own
biochemical activities. In addition, water provides them with
what one can call an air-conditioning system.

Roots also rely on water to hold their shape. The water pressure within them must counterbalance the pressure of the soil
that surrounds them. They must also use water—usually generous quantities of it—as a carrier for the nutrients they collect
from the soil. And the piping of nutrient solutions, initiated by
the roots of plants, is a complicated feat in hydraulics which is
not thoroughly understood. No man-made system of pipelining
or hydraulic engineering can match it for efficiency and dependability. Scientists can speak learnedly in terms of root pressure, osmosis, diffusion, concentration, transpiration, and tension,

but the total procedure involved in the rise of water, along with essential mineral nutrients, from the roots to the topmost leaves is still—at least in great part—a profound mystery.

Quite observably, root pressure varies with the type of plant. For shade trees it is fairly low; for a grapevine it is so high that if one pinches off a leaf in early spring the sap may squirt out like water from a cracked pipe. In pine trees and other conifers the pressure is so slight that no ordinary gauge can measure it. In any and all cases, the development of pressure within the living plant is a function of crucial importance but one that the roots seem to regulate. It varies to suit the needs of plants that range from less than an inch in height to tropical vines a thousand feet long, or sequoia redwoods hundreds of feet high.

Most flowering plants, including the grass of our lawns, exhibit substantial root pressure. If you doubt this, take a close look at your grass early on any spring or summer morning. The odds are that those drops of moisture which so beautifully refract the sun's rays are not really dew, regardless of what the poets and lyric writers may say. They are most probably a visible proof of root pressure—surplus fluid that has been pushed up by the grass roots. Dew is comparatively pure water which lowering temperatures produce by condensing airborne vapor so that it forms droplets which fall on the leaves of grass and other plants. The stains of commercial or chemical fertilizers that these drops contain are a dead giveaway. If you want to be thorough in your investigation of them, you can collect some of the so-called dewdrops and analyze them. You will find that they contain solutions of stand-by plant foods, including calcium, phosphorus, and quite probably the sap broth called glutamine which leaves a sort of whitish smear on drying leaves.

Fortunately, thanks to the wonder physiology of roots, most plants do not leak extravagantly. The vascular system of a fairly healthy plant is amazingly competent. By laboriously

breaking through cell walls, the liquid plant nutrients orginally
taken in by the root cells are passed through other membrane
walls to the inner area or xylem of the root. There they are
mixed with various organic molecules, including amino acids
which the roots themselves have compounded. They are then
ready for the upbound delivery.

In most plants, sap flows from the roots to the leaves through
the inner stem structure. More specifically, the water-soluble
substances and most of the water supply move upward in the
core area of both roots and stem. The return route from the
leaves is mainly by way of the inner bark. The upbound water
is moved more easily than the nutrient solutions, mainly because
dead cells can absorb and distribute it. The proof of this can
be found easily.

Take up a bean plant and dunk its roots and stems in alcohol
—practically any kind will do. The roots and the lower stems
will promptly die, but the leaves will go on flourishing for several days. The alcohol loses its toxicity before it reaches the
leaves, and the dead roots and stems keep right on portaging
water to them.

In living root structures, as in the upper parts of plants, the
transport tissues are constantly being replaced. This is another
reason why roots must keep forever growing as long as their
plant is to live.

Johannes van Overbeek, the former University of Utrecht and
California Institute of Technology plant physiologist whom we
met earlier, has developed a simple device to demonstrate how
the roots push upward the mineral nutrients which they collect
from the earth. Using standard laboratory phosphorus (P-32) as
a radioactive tracer, he watches the sap stream carry it from
the roots upward to the leaves, giving the larger leaves more
than it gives the small ones.

To show how the transport is effected through wood, he uses
a small willow branch. In nature this phase of the traffic is

somewhat baffling, because in some areas the liquids flowing upward on the inner side and downward on the outer side of the cambium tend to mingle. To prevent this, he places wax paper between the bark and the wood so that the flow of the radioactive phosphate through the wood and the flow through the bark can be studied separately. Flowing upward, the chemical keeps dutifully to the inner- or wood-tissue route. When drawn by the sap flow into the leaves, it goes into the leaf cells as casually as an uninhibited child gets into the cookie jar, and once there the uninvited element joins in the marvelous biochemistry of the leaf.

Then, in company with sugar manufactured by the leaves, part of the still-radioactive phosphate flows back into the food stream, some of it going from the leaf through the bark to the growing shoots, stems, or fruit branch. A portion goes back to the root, but whatever its specific destination, the down traffic keeps to the outer or bark-side route. A portion of the tracer stays briefly in the roots, then hops another ride up with the sap stream to the leaves. The van Overbeek demonstration shows that the phosphorus circulates through practically the entire body of the plant several times each day. It also proves that the two-way multilane transport system of a living plant ties its functions and parts together and enables the roots, stems, and leaves to cooperate.

Although admittedly there are still some baffling gaps in knowledge of root pressure, the fact is quite evident that its ability to force liquid through the plant is limited. As a rule the pressure force is evident only when there is little or no transpiration, or evaporation, of water from leaves. Thus, in the morning, when evaporation from leaves tends to be highest, the water loss from the leaves and the movement of water through the stem and roots, caused by the tension thus developed, is faster than any movement caused by root pressure. A simple means for confirming this is to select a smooth-barked tree and place a flexible steel band snugly around its trunk. Then at one end of the band place a micrometer to measure its ever-changing circumference. After the leaves begin to lose moisture to the sun, the trunk of the tree begins to contract, but from then on, pressures and counterpressures effected by the roots begin to register and the sap pressure goes up and down, leaving one to suspect the micrometer of playing some kind of a whimsical, people-confusing prank.

At least for the lay experimenter, as this writer is, a more engaging and amazing game lies in studying the feats in living, working chemistry that roots are forever performing. Unquestionably their initial chemical masterpiece is the conversion of inorganic nitrogen, taken from the free air in the soil, into amino acids, those essential, much talked-of organic compounds that are called the building blocks for proteins.

With only a trifle of encouragement, some roots can produce chlorophyll with great effectiveness. The carrot is an example. In Holland and other places where gardening is particularly imaginative and skilled, and where many vegetables are grown the year round in greenhouses, commercially grown carrots now contain enough chlorophyll to qualify them as a green as well as a yellow vegetable. The productive technique is simply to raise the partially grown root from the soil and permit the upper two thirds or so to turn green in the sun. Expert breeding has produced varieties that are exceptionally sweet, juicy, tender, and (with the added value of the green portion) superior as salad or slicing vegetables.

Plant scientists are increasingly certain that, as far as the basic need for nitrogen is concerned, roots can be made independent of their leaves as long as they are given supplementary feeding. This should include sugar to provide carbon and energy, a few vitamins (particularly B and that member of the vitamin-B complex called niacin), and some inorganic nitrogen which can still be converted to the organic nitrogen compounds they require. On this subject, Johannes van Overbeek observes:

In terms of the alkaloids, proteins, amino acids, and other organic nitrogen compounds, roots are simply not dependent upon leaves. Even when artificially fed, roots can make their own proteins for use in new cell growth, and various other compounds that are essential . . . including nucleic acid.

These activities, it is granted, require the energy that comes from sugar, and respiration, which involves the intake of oxygen. The experimental value of cutting the roots from their top structure is to appraise realistically the actual and potential functions of roots and their capacity to sustain themselves with little or no help from above ground.

The fact that in many known and proved instances roots can be induced or encouraged to get along without leaves opens a vista of vastly improved root crops. Growers are now trying to breed up roots and breed down the tops. Many of the edible roots found in the wild have very slight foliage. A good and widespread example is the common wild sweet potato, which frequently reveals its highly edible underground food root, weighing ten pounds or more, with a top structure no bigger than an infant cabbage plant. A great many perennial root systems can survive for long periods without any leaf support at all. Yet regardless of the great catalog of evidence of how astonishingly provident plant roots really are, generation after generation of "natural scientists" continued to think of them as (a) vegetative anchors and (b) "water" suppliers. During the century just passed this primer-school concept has moved forward step by step, and leap by leap. We know now that the root system provides the stem-leaf portion with most of the water it requires, all of its minerals, many of its growth regulators, and a very great part of the organic nitrogen compounds on which plants and animals depend so profoundly for day-to-day survival.

We are now quite certain, too, that roots sustain beginning plants and dormant perennials until their leaves are somewhere near half grown and otherwise able to contribute to the food supply. Of course, it is no more than fair to concede that once in operation leaves are brilliantly productive. We gather that one of their first pay loads is a vegetative energy compound called ATP (adenosine triphosphate). This is created in the small green bodies, or chloroplasts, of the leaf cell, where the energy of light is somehow changed to chemical energy. The ever-remarkable ATP is the special energy factor for producing amino acids and proteins within the leaf. But the more definitive fact holds that without the substances supplied by the roots, without the figurative engine-starting and the numerous other life-bearing actions and procedures of roots, these leaf "engines" simply could not function and, from appearances, could not— as phrased in American automotive language—"kick off."

One entirely expectable cerebral response to all the foregoing is that one after another of the fields of science is disposed to take over roots as part of its very own. With good cause, chemistry remains the most insistent claimant. As the chemist sees it, the most distinctive function of roots is to gather in the minerals that the plant needs for growth. The effective gardener or farmer gets directly to the practical fundamentals. He knows or soon learns, for example, that potassium is a prime element of plant food. And good crop soil—practically any plant soil—must have potassium, and a very special function of the roots is to collect it. Proof of how hard and how well they work at gathering this essential is shown by the fact that in any working root the potassium concentration averages at least a hundred times that of the surrounding soil. The gathering in of potassium requires a great deal of energy, and roots apparently derive that energy mainly from the old familiar staples, oxygen and sugar.

The magnificent ability of roots as collectors, manufacturers, and compounders of plant foods is never-endingly wonderful and tantalizing. The skill they display involves working associations with vast numbers of other soil lives, including fungi, bacteria, and viruses. But one gathers that the biological and chemical associations simply cannot be separated.

The life-sustaining gist here is that the physiology and the chemical functions of roots are inextricably tangled; roots must take life sustainers from the soil and simultaneously from the leaf, the stem, and the sun. For their own sustenance and for the sustenance of others, they must supply the basic nitrogen both plants and animals must have, and must also gather supplementary elements, or micronutrients, required for vegetative growth and health. "Micronutrient" is a recently coined word for a material that is essential for sustaining life but is not needed in quantity. The lists of micronutrients for plants and animals are fairly much the same. Of the first ten, only boron is required by plants alone. The others list leaders—zinc, vanadium, cobalt, copper, iron, manganese, molybdenum, chlorine, and sodium—are life necessities in both the vegetable and animal kingdoms. The metallic elements are essential as catalysts for manufacturing proteins and other all-necessary compounds; others serve as electronic carriers in the plants' use of light.

Just how important a micronutrient can be was demonstrated recently in Australia, where vast areas of land were once unproductive to such an extent that they were classified as deserts. Yet in terms of actual rainfall these lands weren't deserts at all. When scientists got the problem squared away, they found out that the soil was completely lacking in the element molybdenum, the catalyst required nitrogen fixation by means of bacteria. One ounce of it, properly distributed, will activate an acre of land for several years.

Australian farmers and ranchers are now quite literally making hay as the result of this finding. Fertilizers containing the needed molybdenum were spread on the misnamed desertlands; today clover and other forage legumes are thriving where they had never been seen before. Literally millions of unproductive acres are now burgeoning with farms and greening pastures as the roots, with just a little help from perceptive and patient men, get on with their important work, and Australia scores some of the most impressive gains now current in world agronomy.

To this point we have been shedding coats and getting seated or otherwise readied for the curtain to rise on the greatest show never yet completely seen by man.

Here, while I am sneaking back to return the snitched strawberry plant to its home bed, would be a good place to look back and reflect at least briefly on the beginnings and earlier times of roots. Perhaps more than any other life factor they are causing yesterday and tomorrow at least to hold hands.

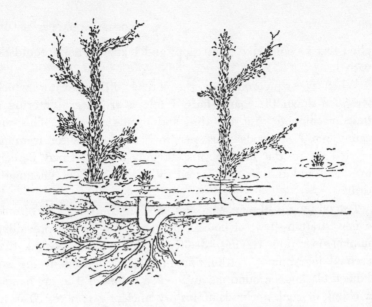

ROOTS OF YESTERYEAR

The roots of plants, my serious young host was reminding me, link yesterday to tomorrow in an ever-provident lifeline. They record the history of ancient eons and they will shape the history of the future.

He had invited me to visit his coal mine near Pittsburgh. During the hair-raising drive to the site he assured me that his mine was very modern and forward-looking but that its story went back about 300 million years to an age which is now called the Pennsylvanian period. Not every state, he added as we skidded around a hairpin curve, has a geological period named after it—as Pennsylvania does. Perhaps less than discreetly I reminded him that Mississippi, or, according to some, the Mississippi Basin, also has a geological period named after it. He replied that I could take Mississippi and I knew what I could do with it.

When we squeezed into the lurchy little elevator which wriggled down the main shaft, I felt as if I were entering a time machine to visit the dim and mysterious past. This uncanny sense of rendezvous persisted as the midget tramline carried us to the farther pits, though lights burned fiercely overhead and the air was jarred by the staccato of pneumatic drills.

I stood by a drill man who was knocking into what he termed a new stratum—new, of course, only in terms of the prevailing mining operation. Having adroitly evaded a rolling chunk that seemed intent upon settling on my feet, I looked at the solidified blackness around me and took a chunk of it in my hand— a chunk of coal, a chunk of history anciently remote. Near its surface I discerned striations of a lighter color which suggested that it might have originated from leaf debris, densely compressed. Noticing my interest, the drill man pointed to a bed, or understratum, of heavy reddish clay which looked as if it had once been a layer of active soil. Here, with the aid of a borrowed flashlight, I found visible indications of plant roots. Many were too vague to identify with certainty, but I soon found evidence of what surely had once been a well-formed taproot at least ten feet long.

Later, while waiting for the elevator, I became aware of another visitor to the mine—a denim-clad young man with almost cherry-red hair. After bidding me good morning, he gave his name as Harrigan and said that he was a mining engineer from Australia.

Australia also has coal mines, Harrigan assured me—not necessarily like those of Pennsylvania, and certainly not all dating back to the Pennsylvanian period of geological time. Some, he said, are "newer" and actually, speaking practically, as an Australian is oftentimes obliged to do, a great deal better. He was mildly interested in the soil-and-roots relics, but he intimated that these, too, could be found even more plentifully and revealingly in Australia.

Next he advised me, with proper verbal footnotes, that the oldest absolute proof of land vegetation occurs in the Silurian rock of Australia, and that the Silurian period of the Paleozoic embraced a time span of about fifty-three millions years, between 405 and 458 million years ago. These land plants included at least two rooted species whose descendants survive today. One of these, he explained didactically, is commonly known as the horsetail.

When I told him that the old brook road to my home in the Vermont hills is practically lined with horsetails, the visitor from Down Under quickly shifted the conversation to the other rock-calendared genus. This was the ancestor of what we know today as club mosses or lycopods, which also grow in the woods around my home in Windham County, Vermont.

The Australian visitor cautiously conceded that club mosses might still be thriving in Vermont, but none of them, he said, could compare even remotely with the ones that had grown eons ago in Australia. His eager blue eyes brightened and, even under the harsh floodlights, showed a gleam of fond recall.

"In Australia," he said, "the prehistoric vegetation was already formalized at that early time. Certainly, sir, our Silurian club mosses had roots very close to a half billion years ago. We don't frankly know much about them except that they were really 100 per cent; I mean by that a very good show . . . Yes, sir, some of those lycopods were trees—some without branches, some with only a few divisions, and some mere stubs standing perhaps twenty feet above the ground. But the trunk diameters must have measured six feet or more. There were others that grew up to 200 feet tall and that without a single branch. Some had great needle-like leaves which practically covered the trunks —they were green tongues, maybe half a foot long.

"About the roots? Really, old chap, you do keep harping on *roots*. You can't really tell about roots until you know your total

vegetation and your rhizospheric environments, most particularly
your geology. What you bloody would-be naturalists need to
do is to learn to project into time."

The informative young man from Australia had something
there. With roots, as with other prime factors of life, it is difficult to know the present or predict the future without at least
some awareness of the past.

Roots have been in existence a very long time; one of the
educated guesses is half a billion years. This guess is supported
by numerous clues and testing devices, including those which
involve the use of atomic tracers, designed to appraise residues
that date back to eras before life forms contained enough mineral hardness to endure as fossils. Paleobotanists are generally if
somewhat vaguely agreed that land vegetation has existed for
at least a billion years—or at least double the span of plant
roots. The oldest algae are believed to be somewhere between
two and three billion years old. It is more or less common deduction that for the first several hundred million years the
growths, at least in part, may have been comparable to primitive algal forms which were neither true plants nor true animals.
It is rather generally though certainly not wholly agreed that
life on earth began in the water, and that adaptation to land
was accomplished by very gradual mutations through millions
of years.

Thus, if one chooses to assume that planet Earth began some
4.4 billion years ago, which is one of the more widely accepted
of prevailing guesses, vegetation may have begun in the vague
and steamy atmosphere of the late Archeozoic period, quite
possibly a billion years ago. But historical geology still surmises
that root structures did not materialize in forms recognizably
like those of the present until what is termed the Cambrian
period of the still very dim Paleozoic period, which ended
about 200 million years ago. This eighty-million-year span which
launched the Paleozoic, or Age of Rock, may have marked the
time when vegetation first succeeded in anchoring itself to the

land, or under very shallow water, with the nurturing assistance of more rudimentary roots.

Thus far, we know with certainty only a little about Cambrian animal life and even less about its vegetation, but the Silurian period, which began about 450 millions years ago, left us absolute proof that land plants, however primitive by present-day ratings, had surely come into existence. As the comparatively brief Silurian period merged into the Devonian (about fifty million years later) fossils tell that first vascular plants—those with veins and conductive tissues—emerged. Many great plant families then came into existence and these were destined to be well preserved in fossil forms or in descendant species which live today. In Devonian rocks of Scotland, Eastern Canada, upper New England and the intermountain region of the Western United States, and also Central Europe, clearly defined fossil remains of rootlike structures have been found.

Included among them are rhizomes, more or less horizontal underground parts of the leafless plant called *Psilophyton*. We still have a great many rhizomed plants, including canna lilies and bananas. The *Psilophyton* was a rather small plant, widely distributed and, from fossil appearances, very long-lived. It ranged in height from eight inches to two or three feet, and it had a woody stem about half an inch thick. The tips of its branches were coiled like emerging fern leaves, and at least some of them bore small pods containing seed spores. The stems had breathing pores, somewhat similar to stomata (the microscopic openings on the lower surface of leaves of present-day plants), and the fossil cell structure indicates a vein system for carrying plant-food solutions from both underground and above-ground sources.

The widespread and somewhat varying fossil evidence indicates that the rhizome or rootstock of *Psilophyton* was more or less parallel to the surface of the soil, and that from this primary runner there probably grew hairlike rhizoids, suggesting the feeder roots of today's higher plants. In some instances there is a recognizable similarity to the top-and-bottom development of certain algal types of water plants still in existence, but the transition to durable establishment in presumably moist land is intriguing. Here, certainly, is credible proof that ingenious and efficient roots were growing in soil more than a third of a billion years ago. This reiterates that although our story is not new, it is certainly changing and improving. There is also evidence that the beginning of land roots and land-venturing animals was more or less contemporaneous.

To this Devonian period (named for England's Devonshire, where some of its more revealing fossil remains have been unearthed) we are indebted for knowledge of what may well have been plant life that then thrived on much of the earth. The Devonian vegetation appears to have consisted mainly of the so-called scale trees, with close-set primitive leaves covering both trunks and branches. In any case, roots would have been needed to anchor the scale trees and there were roots. In an ancient peat bog in Scotland, called Rhynie Chert, a silicified Devonian marsh has been preserved almost intact, showing a colony of plants which lived on (and in) the earth at least 350 million years ago.

Included in this colony is *Rhynia,* a perennial plant completely without leaves. Except for a thick, corklike bark and spore cases which grew on the tips of some of its topmost shoots, this plant was wholly naked. But *Rhynia* had a root, clublike and blunt ended, somewhat larger in diameter than the basal stem and about as long as the plant was high (nine to fourteen inches). This root was placed strategically to tether the plant in the soft mud below the water line. Somehow, apparently without the aid of free air, this clublike root sustained its plant.

Found in this same bog were fossils of the spore-bearing "feathered tree" called *Asteroxylon.* Its trunk and branches, except for a few branch tips which bore spore caps or sporangia, were feathered lightly with short, scaly leaves. Like the naked tree, the feathered tree apparently grew in shallow water and was anchored to the underlying muck by a primitive branching structure which suggests a multiple taproot. Anatomically speaking, in that ancient day the roots and stems of plants were scarcely distinguishable. Even though the anchoring body of *Asteroxylon* is said to have been a rhizome (which, technically, is a modification of a stem), its manner of branching and descending deep into the soil was prophetic of the true roots of today. Quite evidently this was an effective anchor, but it must have been more than that. In some way which cannot be interpreted in terms of present-day vegetation, it apparently helped to absorb nutrients to feed the tree.

When these plants grew, most of the earth's surface was presumably still under shallow water, and a very special challenge for some vegetation was that of growing from the soil through the water into the air. There was the sun's energy to draw on. Beyond reasonable doubt, too, there was an earth's atmosphere with free oxygen, carbon dioxide, ozone, electric energy, and rain. And under these conditions, there developed the miracle of roots.

Perhaps the long, long wait had been for sustenance. There is clear, strong evidence that for billions of years, by present-day estimates at least four billion years, before the Devonian period the earth had been evolving the beginnings of root-available chemicals. Among these were certainly phosphates, manganese, iron, and zinc, and, still more basic, the "Big Four" of organisms—hydrogen, oxygen, nitrogen, and carbon. And there was also ground water needed to carry them. Previously on earth, so one gathers, there had been no indispensable functional need for roots. The early Paleozoic period was a time of marine lives, and there apparently were still no strongly contrasting climatic zones. The books of the rocks tell that the shallow-water plants of arctic lands were then generally similar to those that grew on the southern coast of what is now the United States.

But succeeding tens of millions of Paleozoic years marked the continuing development of soils and chemical salts, which along with water are the prime necessities of root life.

Again reading from the testimony of fossils—or more aptly, the book of rocks—we find that at the close of the Devonian period, for reasons that were partly climatic and partly chemical, a mighty era of vegetation arose. From lower Europe to Australia, existing and ever-multiplying fossil testimony speaks of a great many strange and wondrous plants, most of them swamp-growing ancestors of ferns, mosses, rushes, and—in expanding areas—real trees such as we now list as softwood conifers. Apparently they began as small perennials and gradually attained a size comparable to that of the medium-sized shade trees and smaller forest trees that we know today.

Most fortunately for the animal kingdom as a whole, eventually to include man, when the age of land vegetation arrived, say a third of a billion years ago, rooted vegetation was destined to endure. Within a comparatively few million years, so fossil history tells, plants ranged from microscopically small water forms to huge land plants, some more than 300 feet tall with trunk diameters of twenty-five feet or more. Unquestionably, roots achieved a size to support their principals. During this time the long-lived perennial plant became dominant; many of the genera or families have survived hundreds of millions of years to the present time.

Of the once vast clan of horsetails and their kin, which we have already noted, at least twenty-five species are known today. These appear to have kept quite closely to their original characteristics, except in name. (It does seem silly to label a plant genus *Equisetum* [horsetail] when it has been on the earth at least a hundred million years longer than the horse.)

If one would like to know what some of the Paleozoic vegetation was like, let him look for horsetails on his next stroll or leisurely auto tour through the countryside. These living ancients can be found in wild places almost anywhere from the deep tropics to the upper temperate lands. They often grow beside brooks and ponds, but in regions of plentiful rainfall they can also be found away from the water. Some are ten feet or more tall, others only a foot or two; some supplement their ancient-style roots with more contemporary feeder roots.

Equisetum has a lean, svelte look. It consists of a hollow jointed stem with whorls of tiny pointed or wedge-shaped leaves. At the tips of the fertile stalks are cones containing spores, by means of which the plant reproduces. There are a number of kinds, certainly no fewer than twenty; long ago, by fossil evidence or rock books, there were at least 400 species.

As already mentioned, a handsome grove of these strange but graceful aborigines flourishes along the course of a little brook in Southern Vermont where I live. I have placed about a dozen of them in pans of water in my workshop. Although I took them from medium-moist land, all appear to be thriving in water. Clearly the benefactors here are the roots, which in this species are an astonishing combination of contemporary and very ancient forms. The central root, or taproot, is essentially a continuation of the hollow jointed stem. It is entirely lacking in firm wood but is coated on the outside with a film of resin, apparently exuded in a semisolid state. This coating is insoluble in water but dissolves readily in ether. It is not adhesive; it has the feel of very thin cork, but it is not a true bark. The vascular structure is within the tissues of the thin, partly green stem. The capillaries are extremely narrow, and the upward and downward flow of fluids seems to be accomplished in entirely separate layers.

The central root portions are quite long, averaging at least half of the above-ground height of the mature plant, and quite definitely they contain chlorophyll cells; they permanently retain green coloration. Attached to the tubular main root are very dense clusters of fine rootlets. Their placement is rather irregular, but the structure of the feeder roots appears to be in keeping with prevailing patterns.

The ratio of feeder roots to the tube root is extremely high; on a fairly typical six-inch length of main root I count 319 feeder roots. Including branches and subbranches, the total length of the supplementary system is about 447 inches. There appear to be elongated suction cells or hair roots, too, but their placement is quite random. The total structure looks wonderfully simple, yet cellwise it is bound to be enormously complex.

On the same roadside I keep seeing other more distant descendants of Devonian times, and I continue to marvel at their ingenious and varied root structures. Among these are an engaging host of ferns. From a punching and poking acquaintance with fossils (not including botany teachers) I gather that their remote ancestors (not necessarily of botany teachers) were fundamentally similar to the ferns I see growing along my brook.

Ferns, in fact, are among the most revealing of ancient rooted plants; fossils trace their great history through millions of years. Coal, certainly, is one of the valid souvenirs of ferns that grew more than 300 million years ago. If one splits open and carefully examines a block of coal, there is at least a possibility he may find the markings or outlines of fronds or other fern parts. It is common agreement that the eras of coal and petroleum formation were characterized by giant, tree-sized ferns.

These may very well have been among the first earthly vegetation to attain giantism. Certainly by late Devonian times tree ferns grew to heights of eighty to a hundred feet. Apparently they were shallow-water plants and enormously prolific. In present species of tree fern, a single plant "throws" as many as a billion spores a year, and, significantly, its reproduction can still be achieved in water. But, water beds or no, the ancient ferns which grew as high as a ten-story building must certainly have had sturdy roots; there were mighty windstorms in those times and land surfaces were so near to sea level that great, smashing tidal waves repeatedly swept over the lands.

According to geological reckonings the Devonian period, which marked the coming of the naked plants, including feath-

ered trees, giant ferns, and sizable horsetails, lasted for about
sixty million years. It was succeeded by an even longer span,
the Carboniferous period. Geologists usually designate the first
thirty-five million years of this time of giant vegetation as the
Mississippian period—from about 310 to 345 million years ago.
The forty-five-million-year stretch which followed is the Pennsylvanian period, whose fossils are most commonly revealed to
us in coal.

The Carboniferous period was a vast epoch of encroaching
seas and surging land vegetation which, apparently not certain
whether it belonged on land or in water, tended to compromise
by tenanting both. Mountains were rising, soils were forming,
and the earth as a whole stayed warm and moist, though glaciers
were beginning to appear in the Southern Hemisphere. Marine
animals prowled, gigantic insects began to fly. Ferns, mosses,
rushes, and other wetland vegetation thrust skyward.

Yet on the increasing areas of comparative dryland other
plants, with high canopies of forming leaves, began to cast
deepening shadows. The very fact that so few fossils of root
systems have been found testifies to the multiplying strength of
bacterial lives which must have dealt quickly with undersurface
structures. But certainly the wonder film called soil was forming, and as body-building minerals became increasingly available, both vegetable and animal life became more durable.

The continuing development of deeper soils, teeming more
and more with invisible lives, is indicated clearly by fossil records of the Pennsylvanian period. Throughout both hemispheres
fossils and rock strata tell us that the fortune of vegetation
continued to improve. Widely scattered coal deposits show that
vegetation flourished mightily, even though time and time again
shallow seas invaded the land. The immediate point of all this
is that the Pennsylvanian period, which ended some 265 million
years ago, left us our first reasonably clear picture of early roots,
and marked them as earthy institutions many times older than
man.

The coal-forming vegetation stressed the development of well-mineralized woody plants. Fossil tree trunks of this period show no growth rings; this is evidence that there were still no defined seasons and that what we now list as tropic or sub-tropic plant life grew over most of the earth. However, there is evidence that seasons and seasonal changes in vegetation were beginning to materialize. Glaciers slid remorselessly over substantial areas of South America and Australia. Accumulations of gypsum and salt indicate that a great desert was being formed in what are now Utah and Colorado. Mountains were emerging in such diverse locations as Texas and Southern Asia. And root structures, or at least some of them, were beginning to adapt themselves to the gradually materializing seasons. Even so, as far as we can reasonably deduce, most or all plant life remained perennial. Already that appeared to be the natural way.

The Permian period (named for the province of Perm in Northeastern Russia which has yielded some of the most revealing fossils), extending from about 225 million to 265 million years ago, ended the Paleozoic period. During this time span, continents were uplifted and land life gained greater scope. Climates became more and more varied—more contrastingly cold or warm, dry or moist—and plants began to adapt themselves to the changes. There is evidence that roots continued to lengthen and grow stronger as soils became deeper and primitive leaf structures opened wide to the sun. Minerals were also accumulating everywhere. The most impressive were phosphate deposits, but other primary mineral foods of plants were also being formed. (The mineral deposits of the Permian period still supply most of our commercial phosphates, so essential to plant growth.) Thus, better root systems and improved plant nourishment went hand in hand.

The forward surge of vegetation during the 180 million years of the Paleozoic period continued into the Mesozoic period. This era opened with a time of giantism—giant trees, huge sea reptiles, and ninety-foot dinosaurs weighing as much as fifty tons. At the end of its time span of 165 million years, the dinosaurs and sea monsters were gone, flowering plants had multiplied, and mammals had appeared.

At least three great races of trees (two still known by traceable descendants) flourished mightily in the Mesozoic forests. Perhaps the longest-enduring was the ginkgo or maidenhair tree, which may have been the first tree to bear edible (though not always relished) fruit. The ginkgo of today (*Ginkgo biloba*) still has the same finely veined, fan-shaped leaves and somewhat persimmon-like fruit of the ancient tree. Its root structure consists of a taproot, several diagonally inclined side anchors, and branching feeder roots with well-developed root hairs. The roots are beautifully equipped with vascular tissues.

Mighty conifer forests also began to grace the earth. They were not, however, what we call pines. They were trees as tall as 200 feet, with trunk diameters of five to six feet and had dense heads of leaves about an inch wide and from ten to fifteen feet long. The bare seeds grew in catkin-like cones. Fossil evidence indicates that they had well-developed roots, the carrot-shaped taproot having larger feeder roots.

A large group of evergreen softwoods also arose to replace the marsh and shallow-water giants of the Carboniferous period. There probably were at least 500 kinds. Their leaves were scaly or needle-like, resembling those of some of our present-day evergreens. Presumably their roots were similar to those of surviving species, including the giant sequoia or California redwood.

From all continents and from island remnants of former continents have come fossils that tell of these fabulously great forests in which untold millions or billions of trees towered to heights of 300 or 400 feet. Their trunk diameters measured as much as forty feet, and their life span may well have been more than fifty centuries. All these attributes—size, longevity, resistance to wind and changing water tables—attest the development of strong and efficient root systems.

Also flourishing in the mid-Mesozoic period were cycads as tall as medium-sized trees. They had gigantic pinnate leaves growing out of porous trunks. Among the leaves were structures that looked almost like flowers and in time developed wedge-shaped seeds about the size of a grain of corn. These apparent ancestors of the palm are believed to have had a deep root which may have been a combination of tap and anchor root, and large side appendages which served as fluid reservoirs.

The same general era saw the beginning of flowering plants, the angiosperms, which we now list as the higher plants. With these, roots continued their development. The soil areas surrounding them, called rhizospheres, were well stocked with chemical nutrients and teemed with minute soil lives.

According to fossil testimony, the birth date of flowering plants falls in the Jurassic period of the Mesozoic. The proof—at least until geologists and paleobotanists find a better one—lies in pollen grains, or fragments first discovered in Jurassic coal strata of Scottish mires. More or less simultaneously and from various places in England and Greenland (once land-connected neighbors) have come convincing fossil records of what we rather blandly term "sophisticated" leaf structures. Any student of vegetation quickly learns to read and correlate the root and leaf stories.

The closing period of the Mesozoic period was the Cretaceous (Chalk) period. It lasted some sixty-six million years and was marked by the traceable beginnings of most of the plants that can be seen today, including most of our food plants. One after another, great genera of flowering plants came into being, and, apparently for the first time, the feeble little ground clingers that were to become the grasses began to move out of the deeply shaded tropical jungles. Races of deciduous trees, including the elm, maple, poplar, and oak, became commonplace, and multitudes of flowering shrubs also appeared.

There is evidence that, as drylands and wetlands were more sharply divided, root life continued to become more active and varied, particularly in the drier lands. In Australia, for example, where an ice age was already developing, tree roots probed deeper and deeper into the soil. Rocks and fossils tell us that figs, breadfruit, and other fruiting plants now listed as subtropic were thriving in Alaska and Greenland, but seasons continued to materialize. Winter was touching many of the higher shelf-lands and mountaintops. There, particularly, root structures became more effective and, as one can readily deduce, more intricate in form and function. Cretaceous rock provides fair evidence of the beginnings of root hairs—if not real root hairs, certainly more developed types of feeder rootlets.

As a great group, the early flowering plants profoundly influenced Mesozoic animal and plant life and all life that followed. They shaped the ancestors of our grain and forage grasses and provided food for the grazing animals that were presently to emerge. They prepared the way for our palms and lilies and for our garden, field, orchard, and forest crops.

The Cenozoic (New Life) period followed the Mesozoic and has lasted about seventy-two million years. Its beginning was only yesterday in the history of earth's life. The new life forms that were developed were impressively mammalian—including the most competitive, and now the most numerous, of the higher animals, *Homo sapiens*, man. In remarkable correlation with man, plant life continued to develop and become more and more sophisticated.

Perhaps appropriately, the New Life period began violently, with tremendous earthquakes and volcanic eruptions. Mountains were raised; oceans advanced and retreated. Replacing the dinosaurs and their clumsy companions were browsing animals, and birds more like those of today. The browsing animals crowded into the lush feeding grounds, and living plants, from roots to leaves, tended to show (for want of a better word) modernity.

As the Paleocene epoch, earliest part of the Cenozoic, inched during millions of years into the Pleistocene period, climatic contrasts were increasingly marked. At the beginning of the era, warm weather predominated almost to the poles. And particularly in the Northern Hemisphere, wetlands began drying.

Then, some two million years ago, glaciation began and with it the stabilization of temperate-zone plant life. Ice floes presently covered up to a quarter of the land, ground down the above-sea-level averages of many acres by several hundred feet, drastically revised shorelines, and increased the velocity of rivers and the pace of erosion. Yet despite the thrust of often abrupt and violent changes, the vegetative pattern endured with astonishing strength and tenacity. Almost certainly a prime reason was found in roots.

Thanks in great part to interested amateurs, from high-school freshmen to bulldozer operators, we have splendid fossil records of the Cenozoic. The most significant of all the developments was the gradual adaptation of both plants and animals to their particular environment.

Vegetative roots were highly important factors in the unending saga of adaptation. One after another, the great races of flowering plants, sustained by valiantly adaptable root systems, began to take over the soils of the earth which they had done so much to build and aerate. From polar fringes to the equator, from valleys below sea level to mile-high mountain peaks, from swamps and marshes, riverbanks and ocean margins, to seemingly unconquerable deserts, roots pressed ever deeper into the soils and subsoils. Their adaptability was proved in many ways. Some sustained forest trees skyscraper-high which may have lived for tens of thousands of years. Other flowering plants of the era were smaller than pinheads. Some of the midget species also left contemporary descendants.

A stroll through the countryside almost anywhere will reveal descendants of great timber trees which the ice sheets pushed hundreds or thousands of miles from their original homes. The southern pines, balsams, firs, and tamaracks of the United States are typical examples. They found homes in new places where they still endure. At least as far north as Toronto, any observant stroller can still find in clay or rock deposits indisputable evidence of tropical and subtropical plants. Through many thousands of years, ice sheets, wind currents, creeks, rivers, seas, oceans, and tides have carried plants to other parts of the earth to be anchored again by their roots and re-established as enduring so-called native species. The Cenozoic period particularly demonstrated the amazing enterprise of roots as adapters of lives to places.

If there is no moral involved, and it seems to be a very good bet that there isn't, there is a great deal of satisfaction in reflecting that roots, certainly among the oldest living things now on earth, remain unendingly new. New, not only in the never-ending challenge to know them better, but in their wonderful ability to adapt to specific places, changing climates and environments of today and, no doubt, tomorrow.

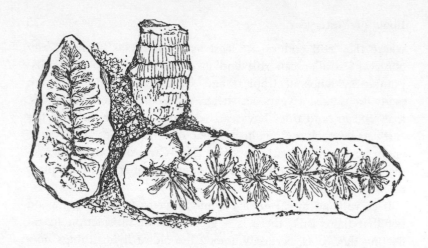

SCHOLAR WITH PICK AND SHOVEL

Back in 1915 an usually quiet and deeply meditative college teacher—"perfesser" was his unasked-for title—called at a bank in Lincoln, Nebraska, to discuss the prospects for a small loan. John Ernest Weaver confided to the bank president that he had some important digging to do, somewhat more than his own time and resources permitted. No sir, he was not digging for buried treasure; he was digging for roots.

The banker eyed his caller with chilly caution. "I sure can't take the likes of that as collateral. I recommend you take it up with the university people. They go for the odd items. I only run a bank."

As a matter of fact, Weaver had already been to "the university people" with what he had described somewhat unfortunately as "an ecology project." At best, the Dean's committee showed no more than an extremely dim acquaintance with the meaning of the term "ecology"; some were not even aware that the newcomer was the recently hired instructor in ecology. But the university treasurer was distinctly aware that at the time the prairie-side seat of learning was fresh out of hard-to-come-by money for squandering on flighty ideas of its extremely new employee.

John Weaver patiently explained that his faculty colleagues were wasting their time trying to learn about plants without knowing anything about their understructures; that the prevailing practice of seeking to improve crops and agronomies by heeding only their top parts was even more futile than trying to diagnose and remedy human ills solely by looking at and "doctoring" only the patient's head or face.

Later, at a faculty meeting, he had tried to better explain his concept of what he termed the life science of root systems. Roots, he said, are the foundation upon which all plant science rests. As recorded in a 1920 bulletin of the Carnegie Institution, he continued in this vein:

> A knowledge of root development and distribution, and of root competition under different natural and cultural conditions, is not only of much practical value, but it also finds numerous scientific applications. . . . It is obvious that a knowledge of root habit is essential to the proper use of plants and plant communities. The plant community has integrated all of the environmental factors of its habitat; it is the fundamental response to controlling conditions. The individual root habit and the community root habit especially, together with the more familiar above-ground parts, serve to interpret these environmental conditions. . . .
>
> In classifying lands for forestation, for grazing, or for agriculture, a knowledge of root habit and extent is of prime importance. Indeed, such a study not only forms a basis for judging the natural vegetation as an indicator of the value of lands for crop production, but also for determining the kind of crop

to be most profitably grown. It leads to the intelligent solution
of problems of range improvement, the selection of sites for
forestation, and of numerous other problems where natural
vegetation or crop plants are concerned. The application of
knowledge of the root development of crop plants in the prep-
aration of the seedbed, rate of seeding, method of tillage, use
of fertilizers, irrigation, and crop rotation are too patent to
need further discussion.

Whether or not his academic superiors listened, John Weaver
spoke on, confident that in time gifted students would listen
and act accordingly. On the basis, and most importantly, John
Weaver spent half a century studying roots and helping others
study them—not theoretically but by actual sight and touch.
This was his special gift, to which he devoted his life. As long
as he could remember, Weaver had been fascinated by the
normally unseen guts of vegetation. On or about his twelfth
birthday he shaped a resolve to discover the root structure of
every kind of plant in his homeland. By his fortieth birthday
he had succeeded in unearthing and viewing the root structures
of about 1500 plant species common to the American Great
Plains. Before and after that milestone he filled and in part
published about twenty volumes of notes, drawings, charts, and
other measurements of living roots.

Fortunately for the world at large, John Weaver spent most
of his life in and around Lincoln, Nebraska. Within easy access
were, and are, some of the most revealing root beds to be
found anywhere on earth. For here are vegetative communities,
including the great Gramineae—tall, short, or intermixed perennial grasses which for uncountable centuries have followed
world-significant vegetative patterns. Over these lands probably
for millions of years have walked the echelons of grazing animals
which helped so memorably with providing fertile lands for
man. On these Great Plains grazed the forebearers and earlier

legions of camels and horses and the ancient earth-circling elephant kinds. Here more recently grazed the mighty herds of Western bison and the renegade longhorn cattle, which some students believe numbered in tens of millions of head and perhaps included some of the largest herds that ever walked the earth.

Here is more or less centered one of the oldest and best-tried true prairies to be found anywhere. Included is "tall-grass country," with rainfall averaging around twenty-eight inches per year and grass roots which reach down to average depths of eight or more feet. Included, too, are the "short-grass plains," with rainfall averaging no more than fourteen inches per year and preponderant populations of short or tufted grasses with more shallow roots deftly clustered to make the best-possible use of occasional summer rains. All present, and on all sides, are, or at least have been, "mixed prairies" where short grasses and sedges form a kind of oversized lower layer for the taller prairie species. Scattered more generously than one at first glance might note are various shrubs and trees and other woody plants, and present, too, are various great cultivated crops adapted, though not always competently, to the many types of prairie soils. Thus, Nebraska, as a far-flung whole, is a world museum of typical root structures. John Weaver recognized this fact and contributed momentously to making it the seed of world-wide knowledge.

In turning through the various logs and journals of John Weaver's lifelong exploration of roots, one can pick at random hundreds of revealing generalities. First off, one notes that by nature's will there are very few shallow root structures in a true prairie. Practically all indigenous grains extend into the fifth foot of soil; a majority of the perennial species have roots that penetrate eight to twelve feet; a few reach down twenty or more feet. The lateral spread of such roots is rarely more than four inches, usually less.

By the time John Weaver was a high-school sophomore, he had developed his own special technique for digging out root systems intact. Having chosen a plant that interested him—and practically all plants did—he would dig a trench beside it about five feet deep and wide enough to permit him to maneuver. That done, he would settle down to really serious digging. He had devised a pick with a special cutting edge. With that he would "open a face." Then, with the patience of an archeologist, he would excavate the root structures.

"This apparently simple process," he explained, "requires much practice, not a little patience, and wide experience with soil texture. To assure certainty as to the maximum depth of the root termini, the soil underlying the deepest root is usually undercut for about a foot below the root ends, and is carefully examined as it is removed."

It is outrageously hard work which by its unalterable ground rules permits no mechanical aids or labor savers. Every stab of the shovel or swing of the pick or facing adz must be suited to the placement of the roots and dependent on sharp eyes and firm muscles. Most of the time one waters his way along with his own sweat. John Weaver described a typical plains or prairie grass as a species having a top portion sized to fill a hat and a root structure sized to fill a rain barrel. With his special genius for self-belittlement, he once described his role to me.

"A grass-root excavator," he said, "is a special brand of dang fool who is willing to dig down anywhere from ten to thirty feet, and through a like number of centuries, to take out a root system complete. . . . He may strike water often enough to qualify as a well digger. Not long ago I brought in a gusher of a water well just as I got to the lowest root cap, thirty-one feet below the surface, of an eight-year-old alfalfa plant.

"I always say a man with a back strong enough to get to the bottom of any root system that's important—and just name me one that *isn't* important—deserves a rating as an ecologist, or maybe some other kind of four-syllable crackpot."

The need for a strong back gave due acknowledgment of the resistance of Great Plains earth as well as the strong, sure reach of its enduring and sustaining roots. Most soils throughout the American Great Plains are of hard, compact strata, frequently including so-called impervious hardpan. But even the hardest and longest of these are definitely accessible to roots, beginning with grass roots. For example, in the hard drylands around Colorado Springs and in western South Dakota, excavations of the root structures of thirty-six typical plants show that only four, including two cacti, are rooted in less than two feet of topsoil. Sixteen species, including ten grasses and sedges, reach depths of at least five feet; while the remaining sixteen, including three grasses, grow down from seven to thirteen feet.

In the mixed prairies of the Colorado and Nebraska Sand Hills, John Weaver excavated and described forty-five plant species that he considered typical of the enduring native vegetation. Of these, only four have root systems preponderantly in the topsoil. The pick-and-shovel scholar continued to learn about and record with words and pictures the amazing talent of plant species to fit their root structures to the special needs presented by various growth sites and the depth of the below-surface water sources. He learned from hard and careful digging that on hilltops and hillsides the root lengths of identical species tend to exceed those that grow in the valleys. Particularly in sandy soils and hard-packed loams, the native vegetation usually provides itself with rather widespread lateral roots within ready rain-catch distance of the surface; this is specifically true for four fifths of the sand- or clay-grown species that John Weaver examined.

Wisely, the root scholar also directed his root explorations to the principal cultivated plants, beginning with the chief grains. His first impressive finding was that even the annual crop plants have much deeper root systems than is commonly supposed; none is completely limited to the upper soil. Even the common cereal grains, wheat, oats, and rye, which tend to form clumps or clusters of feeder roots in the topsoil, have main roots that penetrate the lower soil strata, including the hardpan, to depths of five to seven feet.

For example, on fairly typical prairie-type lands near Central City, Nebraska, John Weaver proved that the average above-ground height of the common wheat plantings was a shade more than thirty-nine inches, and the average root depth was eighty-four inches. In very hard clay in the area of Lincoln, Nebraska, he found that the average plowing depth was four inches. Despite this and the tough substratum of heavy clay, the average root depth of wheat was forty-five inches. Oat plants in the same area grew to about three feet tall, but their roots penetrated the very heavy and resistant subsoil to a minimum depth of four feet, frequently more. He found that instead of being discouraged or frustrated by the hardpan clay, the roots of corn plants pierced through it until their feeder roots touched depths of about nine feet.

Regardless of extremely hard subsoils, a common cane or grass sorghum regularly establishes root depths of more than four and a half feet within seventy-five days after planting. Root structures of the sorghum kin, kaffir corn and milo maize, frequently descend six feet during an average growing season. Interestingly, too, these and other long-acclimated dryland crops, including commercial sunflowers, supplement their vertical root growths with robust lateral and topsoil roots which serve as rain catchers for growing season precipitation—when and if it comes.

John Weaver wrote in conclusion: "The great depths reached by crop plants under the true prairie environment indicate that they, like many native species, must rely largely on the deeper portions of their root systems for water and mineral solutions when approaching maturity and especially during periods of drought. But once well begun, the roots will take care . . ."

The scholar of roots continued to point out that the root growth effected by annual plants is similarly the incessant

miracle of nature and, for good measure, a demonstration of
temporary building of quality standards which could endure
heaven knows how long. In the course of growing seasons that
range from sixty-six days to 116 days, a typical field crop such
as corn, wheat, oats, or rye builds a root system potentially
capable of functioning for a human lifetime, or quite probably
for a century or longer. Even when bereft of any chance of
permanency, however well deserved, a typical annual root seeks
to keep growing from first seed dropping to onset of hard-freez-
ing winter. It does not always succeed adequately, as we shall
presently note, but nobody can doubt that it tries ever so
devotedly. Inevitably, hardpan or similarly recalcitrant under-
layers reduce the down growth but almost never halts it. Perennial roots, with the many advantages of that fortunate state,
are also slowed by impacted substrata, but as a rule considerably less so than the annual roots, in great part because once
on the build and get they continue to grow during winters.

Understandably, Weaver's untiring interests in roots and en-
ergies and skills for exhuming them gravitated to perennial
plants, particularly the grasses and prairie shrubs. "I would
cherish to work with tree roots," he wrote a colleague, "but being
a natural-born sodbuster, I just can't seem to get far away from
grasses and their companions." As a ready compromise he began
a long and brilliant study of alfalfa, nowadays the most frequently grown legume companion of grass. Here, truly, is a
perennial miracle. In his customarily mud-smeared report book,
the pre-eminent root scholar celebrated his fiftieth birthday by
compiling what he termed a root journal for alfalfa:

Alfalfa: Area—Lincoln, Nebraska. At Age of Plant as Follows:

Age of Plant	Height of Top	Depth of Root
38 days	0.3 foot	1.4–1.7 feet
75 days	0.9 foot	4.3 feet
116 days	1.1 feet	6.3 feet

Six years later he scrawled at the lower margin of the same
page:

"Six-year-old plants rooted in stiff clay penetrate average depth of 12.3 feet." Two years later, John Weaver jotted a final postscript:

"Eight-year-old alfalfa roots through stiff hardpan undercover to est. average depths of 15 feet, maximum 31 feet."

Throughout a long career of studying and discovering plant roots and encouraging others to go forth and do likewise, John Weaver won a remarkable variety of followers. These included farmers, gardeners, livestock raisers, and—in the more formal fields—pathologists, soil chemists, botanists, physiologists, ecologists, and, of epochal consequence, vegetative geneticists, many of whom prefer to be listed simply as plant breeders.

The scientific community and interrelationships that the pick-and-shovel scholar helped draw together or literally establish keep growing in effectiveness. By the 1920s, Weaver found himself serving quite literally as midwife and nursemaid for the great emerging science called ecology and the engaging and no less brilliantly advancing science complex known as genetics.

John Weaver foresaw this far in advance. Long before any public had begun to take him seriously, the philosophical root digger of Lincoln, Nebraska, foresaw and quietly predicted that the various agricultural sciences—indeed, the entire range of natural sciences—would find their best hope and progress below the surface. This conviction went much further than infatuation with his chosen enterprise. Weaver clearly sensed that learning about roots is indispensable to defenses against the majority of vegetative diseases, including the ruinous hosts of insects, fungal, bacterial, and nematode enemies of vegetation and, inevitably, of people and animal life as a vast whole. Even during the 1920s alert pathologists were beginning to learn that a commanding majority of vegetative diseases are actually root diseases. Vastly improved knowledge of root life is a must for competent defense. Weaver was early convinced that regardless of the deluges of significant findings about human and animal diseases, the vegetable kingdom is actually less healthy than the animal kingdom. Better knowledge of roots is an absolute prerequisite for raising its health levels; and even a slight improvement of root hardiness could greatly increase all kinds of harvests. As a pantheist of sorts, and a lover of gentleness, Weaver reflected that vegetation is the more gentle "world of life," far less predatory and cruel than the animal world.

But quite apart from his more general background of believing and sensing, Weaver was uncovering a vast, specific, profoundly valuable, and engagingly dramatic life sphere in the subsurface "world."

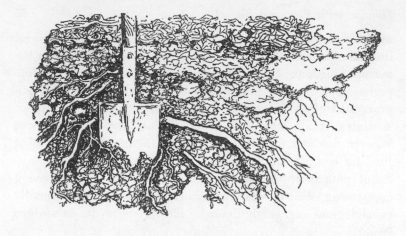

6

THE TABLE OF ROOTS

The Romans had an especially pertinent word for the ways of living roots in living soils. This word has been revived as "commensalism." Its Latin roots (no pun intended) are *cum*—"together"—and *mensa*—"table." This gist of meaning is eating together at a common table. In biology shoptalk "commensal" has come to mean associated with or living together in an intimate, more or less continuous manner or vital relationship.

Stated quite literally, roots and their fellow tenants of the soil share a common table at which they eat together (sometimes they eat each other) and for good measure help serve the foods, clear the table, and attend to the garbage, together. Thus, the concept of soil as a vast extension of dining tables is a meaningful metaphor. Ever in close, life-building association with roots are tinier beings—bacteria, fungi, insects, and others—in numbers so vast that man has never yet been able to count or compute them. Millions upon millions of individuals occupy the soil particles that surround the roots. Many kinds are mutually dependent; others maintain a constantly cooperative relationship with the roots.

Bacteria, which unquestionably are the most abundant of the underground associates of roots and the most valuable of soil builders, are perhaps the earliest of all known plants. Certain kinds, called autotrophic bacteria, have left their traces in rocks of more than a billion years ago. Apparently they lived then, as great numbers do today, by oxidizing iron, carbon, hydrogen, sulfur, nitrogen, and other elements. Later forms came to rely on organic sustenance.

But there are some food elements that bacteria cannot synthesize solely by their own devices. Vitamins such as thiamine, niacin, riboflavin, and others are on the list. For these and other special requirements, great numbers of soil bacteria depend upon the exudates of roots and upon other soil lives. (The soil is ever an institution for both giving and taking.)

Beyond reasonable doubt, bacteria remain the most numerous of all living forms on earth. As yet microbiologists have not been able to see clearly or describe formally more than a tiny fraction of the many species believed to exist, but they have seen enough to know that bacteria are the primary builders of the living soil and therefore the first indispensables of living roots.

Yet so tiny are these one-celled bodies that, according to my own shirt-cuff computation, about a quarter of a million of some of the races could recline on a medium-sized pinhead. Working bacteriologists tell me that when they are questing for bacteria, it is their habit to set their microscopes at a magnification of 1000 and hope to go on from there. They add grimly that any effective measuring device must be marked off in microns. (A micron is one thousandth of a millimeter, or about one twenty-five thousandth of an inch.)

While visiting a University of California laboratory, I was able, with some expert assistance, to get a clear view of what was introduced as a "reasonably typical" bacterium from the soil. This was a kind of subminiature teardrop, practically colorless except for some tiny droplets of a grayish fatty material, enclosed in a gelatin-like capsule. I observed, as many have before me, that some bacteria are equipped with wavy threads, or flagella, which serve as propellers. Others have miniature oars for paddling through microscopic moisture, and many are merely fat and round but with noticeable differences, as with roly-poly people. Some produce "resting spores" by means of which they multiply. Others propagate by routine cell division.

These one-cellers have the potential of reproducing themselves two or three times within an hour. Right now, I am striving to watch the "birth" of an ordinary nonspore bacterium. The cell keeps swelling, elongating, and rounding until its initial size is about double. It becomes constricted at, or near, its center. The constriction deepens until the cell divides in two, the walls of both cells being faultlessly repaired. Now two bacteria are ready-set-go to participate in a population explosion which has been going on presumably for thousands of millions of years.

The potential rate of increase involves astronomical figures. By attainable count, within an hour one bacterium can become four, in three hours, sixteen; in eight, 64,000; in sixteen, about a million; and so on. The best proof lies in the living soil. By broad estimates, one gram, or about one thirtieth of an ounce, of surface soil holds anywhere from one million to fifty million living bacteria, plus other forms which may number at least in thousands.

As with most other forms of life, there are natural controls for keeping the legions of bacteria from getting completely out of hand. Among these controls are the limiting factors of food, water, warmth, and still-mystifying restraints such as the accumulation of bacterial residues.

Among the uncountable multitudes of soil bacteria the spore-forming kind appear to be the more durable or stable type since they tend to be impervious to drought and their longevity is usually greater than that of most drought periods. Some species are said to live for ten years or perhaps longer. Most soil bacteria are, or are believed to be, the spore-forming kind. This is more than a small blessing to toilers with microscopes, because the spores are comparatively clearly defined and much easier to see than the source lives.

Comparatively few of the soil bacteria are to be listed as entirely parasitic. Some, indeed, are completely self-sustaining. Others live by oxidizing chemical elements, or by using the exudates of roots, or by sustaining themselves on other soil lives. Some of the bacteria perform chemical conversions for the common good of the soil community, and otherwise help their fellows grow and multiply. Others serve as restrainers of otherwise inestimable bacterial birthrates, and this, in the soil or above it, can be of great benefit.

Vast numbers of bacteria thrive on dead organic matter, both plant and animal. In the process they break down its components into simpler organic structures that can be absorbed by the roots of plants and used as plant foods. Other bacteria serve invaluably to unite ammonia and carbon dioxide to form amino acids, the "building blocks" of which proteins are made. The importance of bacteria in decomposing dead organic bacterials can hardly be overemphasized. Were it not for their work, essential nutrients would be forever locked away from plants in large unmanageable molecules.

The actual chemistry of bacteria-induced decomposition is not well understood. However, we know that these microscopic craftsmen perform the task with infinite skill. They break down the waste, salvage its components, and return to the soil a vast treasury of food elements. The recoveries include most of the food staples needed by plants.

Another important function of soil bacteria is the recovery

or fixation of nitrogen—the most basic of all food elements. The
main source of nitrogen for animals is the breathable air, but
the animal cell cannot convert air into food. In its raw or
natural state, nitrogen is the wild mustang of the elements,
indisposed to join the remainder of the herd. But bacteria have
ways of making the raw nitrogen of air directly usable. A
common chemical sequence is the production of ammonia, the
changing of ammonia to nitrites, and the changing of nitrites
to soluble nitrates. Apparently each step is handled by a different race of soil bacteria.

A procedure frequently mentioned is that provided for specific root spheres by *Rhizobium* types of bacteria, some of which
take nitrogen from air contained in the upper soil and implant
it in nodules, like little gift packages, on the roots of the
rather diverse group of plants we call legumes—peas, beans,
clovers, and their many-sized relatives. Because man has belatedly learned of the wonderful function of the most effective
of gift and relief agencies, the legumes are being surged forward
as favored crops. Almost overnight, soybeans have become the
number-three field crop of the United States and throughout
much of the Western World. Alfalfa, another legume, is taking
over as the foremost forage crop.

Man discovered, or identified with certainty, the root nodules
on legumes less than a century ago. This was a trifle late.
Paleobotanists are generally agreed that the nodules, and the
bacteria which put them there, have been in the soil for at
least fifty million years.

Even more belatedly, people continue to find more and more
plants possessed of roots with special talents for collecting and
making available to their soils even more generous measures of
nitrates or similar compounds. Included recently in the ranks
of the nitrogen fixers are the Australian pine (*Casuarina*), the
oleaster, alder, sweet gale (*Myrica*), and the mountain lilac
(*Ceanothus*). This, clearly, is added evidence that usable nitrogen which accumulates in the soil is immensely valuable to

vegetation, to the grand conglomerate of soil life, and to the animal world. Without it no soil could long remain fertile, and little if any earthly life could long endure.

Though they have been the indispensable companions of roots for millions of years, the soil bacteria do not work alone. Among their more numerous helpers are the fungi, including the legions that live exclusively on dead matter. They work with the soil bacteria in the task of making the remains of plants and animals available to roots. In this unending resurrection within the living soil, bacteria digest and free proteins—or reduce them to assimilable forms. The fungi mainly digest and break down the complex carbohydrates which include principal portions of deceased plants. As part of this very special function the saphrophytic fungi (those that live on dead matter) also help invaluably in freeing and recovering nitrogen, hydrogen, carbon, and other essential elements.

Like the bacteria, the fungi manage to travel far. Their spores move in the clouds and float in the air high above them. There is no known part of the earth without fungi, and no known land surface too barren for their tenancy. In drylands and deserts, on mountain peaks, in polar lands, and on most islands, there are usually many times as many fungi as seed-bearing plants. It is noteworthy, however, and perhaps hope-inspiring, that on continents and other very large land areas there are usually more different varieties of seed-bearing plants than of fungi.

Many believe that fungoid life forms are fully as old as roots or vascular plant life; some scholars even hold that they are much older. The fossil trails are very far from being clear, but fungus spores have been found with the debris of higher plants that grew a third of a billion years ago, and the characteristic threadlike hyphae have been clearly distinguished in fossils dating back to Devonian times or beyond.

While some live on dead matter, the fungi are primarily parasitic, dependent for their subsistence on other lives. Apparently no fungus is entirely self-sufficient in terms of its food supply. None has chlorophyll, and none is able to synthesize its requirement of carbohydrates directly from carbon dioxide and water, as do green plants in the presence of light.

In size the fungi vary from one-celled microscopic forms to comparatively gigantic growths such as the giant puffballs and other king-size mushrooms. Fungi multiply by spores, in the main, but in some instances by the breaking apart of a portion of the rootlike tangle of filaments known as the mycelium. The soil roles here are of very great importance.

The surface of the mycelial threads absorbs soluble minerals and other nutrients. These diffuse through the cell walls into the protoplasm. Some fungi are able to live entirely on fats, others entirely on ammonia. But the great majority require additional ingredients such as external and ready-made supplies of organic nitrogens, various vitamins, and the amino acids. Usually the mycelium portion secretes enzymes or digestion aids through its walls into the outlying soil or other substance where the fungus is growing. The enzymes act directly on the nutrient fluid, digesting part or most of it outside the fungus and leaving the soluble part to be readily absorbed.

However, great numbers of fungi depend heavily on the giving power of roots, which recover and synthesize a great part of fungus nutrient requirements. The life-sustaining dining table of the soil also has its crumb snatchers and garbage collectors. The air and water requirements of roots are shared by the fungi. As soon as the mycelium has reached a suitable size and stored sufficient food, it, too, in many instances, develops a reproductive body containing spores. As with most other forms of life, there are good fungi and bad. Unfortunately for people and their crops, the criminal proportion is uncomfortably high. Even so, the beneficial fungi are sometimes very beneficial indeed. To note only two examples, there are the tiny single-celled members called yeast, which cause our bread to rise, and in the soil are scores of species of the so-called symbiont fungi, which live with other organisms to their mutual benefit. Many of the soil molds belong in this group.

There are also parasitic fungi that are almost magnificent artists at teaching. Here the lichens are among the most widely prevalent examples; they vie with the mushrooms for honorable mention among the fungi that are directly edible by men and other animals. In the arctic and subarctic, where they are known appreciatively as "reindeer moss," lichens are the great sustainers of reindeer; and lost hunters and trappers have stayed alive on the nourishment obtainable from lichens soup or broth.

In a sense, the lichen is a conspiracy of leechers. It is made up of fungi that cover and tap into algae, with the fungi tending to get the better of the relationship. This very special fungal talent for getting the better of a relationship helps form the wonderful world of vegetative roots. Parasitic fungi now account for an actual majority of what are regarded as the more serious root diseases and above-ground maladies of vegetation. This is one of the more puzzling areas of pathology. The major fungal enemies of vegetation are, in their own bizarre ways, specialists. As a rule, a given fungus attacks or exploits only one or a very few species of plants and, in many instances, limits its residence to a single portion of a plant. In other instances, all too numerous, the injuries from a villain fungus are visible in only one part of a plant. Quite frequently it carries its life cycle much further. One of the best-known examples is commonly known as corn smut. Here the fungus grows and fruits in the tissues of tassels and ears, where it produces large tumor-like swellings. But the fungal invader takes its feed from the stalk, and perpetuates its kind by dropping its spores on or into the soil. In due course the spores germinate in the soil and from that vantage, wait, ready to presently invade and infect the higher portions of the host plant.

The parasitic ways of fungi are enormously varied but oftentimes are disgustingly logical. Many live directly from living tissues of the roots. Some produce different types of spores suited to invading two or more different kinds of plants. One of the better-known examples of this is the common wheat rust which produces two different types of spores while growing on wheat or other grasses, and two other types especially tailored for parasitizing the leaves of the barberry bush.

Fortunately for vegetation as a whole, there are also fungi that live only as parasites on other fungi, which provide a natural and one of the better ways to effect control. This kind-on-kind leeching has many elaborations. There are other parasites that place their mycelia in a way to permit them to take nutrient from a root or other plant organ by way of another fungus or fungal group which is parasitizing the same area. Still other fungi take nourishment directly from the bodies of sucking insects; common scales of fruit trees and grapevines provide examples of this. But many of the damaging relationships between fungi and insects occur definitely by happenchance.

All too often insect parasites take advantage of scars or openings in root tissues caused by fungi and vice versa. In the instances of several serious disease pathogens, parasitic insects when changing locations carry fungus spores with them. But however good or bad its conduct, the fungus, like all things living, must eat and drink. Here the world of the fungi yields to a momentous disadvantage. So far as is surely known, no fungus is capable of producing its own food and drink. In terms of digestive systems, including appropriate enzymes, no life form is more effectively endowed than the fungus. But feeding is something else again. Known species of fungi live entirely on fats or ammonia. But the huge majority require outside and ready-made supplies of organic nitrogens, vitamins, and amino acids. Roots happen to be especially proficient in recovering and synthesizing or otherwise providing what fungi need to eat. And root requirements of air and water are also common to most of the fungi.

It follows that the soil tables that in due course supply our own dining tables are unendingly afflicted with mooching fungi.

As with other kinds of neighbors or kin who wait, ready to eat, by no means all the free loaders are obnoxious. Some, such as the saprobes, which help valuably in recovering all manner of dead and decaying materials, not only are good for roots, but are most valuable friends to the vegetable kingdom at large.

There is no escaping the fact, however, that soil fungi do indeed cause and perpetuate thousands on thousands of root diseases. In the main, as in the case of smuts and rusts, fungus-caused diseases are best known by their above-ground symptoms. But home gardeners share with the most learned pathologists the now absolute knowledge that fungi are most damaging in or on the roots.

Sooner or later almost any home gardener makes the unwelcome acquaintance of the commonplace and conspicuous root rots that destroy beets, turnips, rutabagas, cabbage, cauliflower, and other garden stand-bys. The culprits here are of the insistently antisocial fungus family known as *Plasmodiophora*.

Fungus-caused vegetable diseases are historic as well as spectacular. One that profoundly influenced American population was, of course, the great Irish potato famine which reached its people-saving climax late in 1845. The so-called potato rot or late blight, caused by the fungus *Phytophthora infestans*, apparently moved out from the homelands of the common potato in the upper Andes to reduce severely the number-one food vegetable not only in Ireland but in great areas of Western Europe as well.

Only a few years later, during the 1850s, another fungus disease almost annihilated the then most important coffee industry of Ceylon and much of Southeast Asia. During more recent years other fungus diseases of vegetation have made dramatic and historic appearances. By long and costly experience, people have learned that most of the far-reaching and more ruinous epidemics of plant diseases are fungal.

Americans were graphically awakened to this fact in 1904 when a young and vehement President, Theodore Roosevelt, began to shrill the disturbing news that the "history-ladened" chestnut trees on Long Island and adjacent areas were beginning to die of what T.R. termed a "peculiar blight." Investigations showed that the pathogen was a little-known fungus lately imported from the Middle East. In its homeland the miscreant had not been especially harmful, but once released in America the blight turned virulent and presently destroyed practically all the edible chestnut (*Castanea dentata*) in North America.

Many people, including the writer, remember the intercontinental epidemics of wheat and other cereal rusts that back in 1916 began ravaging grain harvests from Australia to Kansas, from India to Texas, from South Africa to Germany, when three fourths of the entire wheat harvest of that year was erased by the destroying fungi.

Nowadays plant pathologists tend to show less alarm about vegetative contagions and to give more intense study to the enemy fungi that afflict the root structures of so many valuable crops without the spectacular testimony of vegetative epidemics. Among these more persistent fungi villains is the *Phycomycetes* family, root enemies of grains, various other grasses, and leafy vegetables. These enemies include the summer spores or powdery mildews that afflict cereals, clovers, strawberries, gooseberries, roses, apple trees, and even oak trees. There are also the fungoid ruiners, *Schlerotinia*, that invade both the roots and the circulatory system of apples, plums, and other fruits.

The known list of these so-called lower-order pathogens keeps growing; for the most part, the harm-doers concentrate on the roots, where, for better or worse, plant health is so largely decided. While man's knowledge of the fungoid enemies of roots is growing rapidly, it is certainly not new. Back in the fourth century B.C., Aristotle and his naturalist pupil Theophrastus recommended the upbreeding of fungi-resistant varieties of wheat, rye, figs, and other crops that even in those times were direly afflicted by root diseases. In its turn the Old Testament speaks of mildews, "blastings," and many other vegetative diseases caused by fungi, including those of the soil.

During the American Revolution, Abel Cheatham, a professor of natural science at Harvard, listed 100 species of what he termed "fungals." At present approximately 100,000 species of fungi are known or, as the saying goes, authentically described. Of these only about 400, certainly less than half of 1 per cent, are known to be harmful to plant life. But the mycological rogue galleries are still open, beginning with the roots. A recent count by the U. S. Department of Agriculture mycologists established that of 384 species or subspecies of plant-injuring fungi recently studied, 94 per cent (all but twenty-three) are known enemies of roots.

Even so, the most numerous freeloaders at the soil tables are bacteria and insects. The latter, so far as we now know, are far and away the most numerous members of the animal kingdom. The number of insects or closely related species is currently estimated as about 3.3 million. Dr. T. R. Hansberry, who also served Cornell as a professor of insect toxicology, estimates that an attainable count of the insect kind and related genera may now well total as high as 3.75 million species.

By comparison with fungi, insects are newcomers on earth, though apparently nowhere near so new as man. In any case, fossils indicate that insects have lived on earth for at least 300 million years; the fungi much longer. But more insects live in the soil than anywhere else, though water and air have their share. Insects relate to virtually every utility or possession of man, beginning with his body and progressing to his petroleum supply in which certain insects are known to propagate.

But the vast preponderance of insects are not harmful to vegetation. According to prevailing estimates, about 600 insect species can be listed as known root pests, while approximately half that number are known to be harmful to above-ground parts, particularly leaves, stems, flowers, and fruits or grains. It is common knowledge that crop losses caused by insects are formidable; the toll in the United States is still listed as about 10 per cent of commercial harvests. The World Food Organization of the United Nations estimates that the world total of food losses caused by insects is no less than twenty billion dollars annually. This estimate admittedly relates only to harvests. Root systems may very well record a much greater total of insect damages.

The fact stands that the enormously populous and diverse world of insects is also rooted in the soil. Hundreds of thousands of insect species regularly lay their eggs in the soil; hundreds of thousands more use the soil as second homes for pupal or other metamorphoses, or as retreats during winter, or other intervals of dormancy or "dispause."

Inevitably and naturally, the home sphere of roots doubles as a refuge for insects. The reasons are self-evident. Soil is the surest retainer of water, and insects, like fungi and animals in general, cannot endure without water. Soil, too, is the great moderator of temperatures. Even when the air at ground level is twenty degrees below zero, the soil temperatures are rarely below twenty-eight above. Surface air temperatures of 120 degrees rarely find soil temperature higher than 90 F. Most insect eggs are adjusted to relatively moderate temperatures, such as those that most soils maintain. And the same living film is usually the best-attainable nursery and growth ground for insect young. Water, food, and protection of life from the violence and vagaries of climate are of primary appeal to the almost fabulously varied insect kind. It follows that this most varied life form which is capable of damaging every physical possession of mankind is, and from all indications will ever remain, a principal enemy of roots.

As a vast group, the insect is well suited to his place at the soil table. He requires food, water, shelter, and a stability of temperature; and, again as a vast group, he has pre-eminent facilities for attaining what he requires.

Thus, there is reason to believe that the soil tables, enormously crowded as they are, will become even more crowded. The already attained multiplication of the table guests, servers, cleavers, and upsetters is far beyond man's present capacities to count. Bacteria can propagate dozens of times within a day. Multitudes of fungi and insects propagate dozens of times within a month. All must eat and drink. It follows that in the wondrous jungles of fecundity directly beneath our feet there can be no absolute limitation of table places.

But there must be a place and time for separating the good from the evil, the benefactors from the destroyers. Soil scholars are now inclined to agree that odds for benefit from bacteria are at least 10,000 to 1; for soil fungi, about 2300 to 1; for soil-based insects, perhaps 4000 to 1. Yet even the tiny minorities of the bad behavers at soil tables are more and more ominous and imperiling to people.

The same holds true for other orders of underground animals —particularly, as already briefly mentioned, the class or phylum, (or perhaps better say, the tissue animals) called nematodes. The nematode, or threadworm, is ever a remarkable animal; it does not have a developed brain, yet for more centuries than can be surely counted it has been outsmarting man. It lacks a nose, yet it can trail more effectively than any bloodhound. It has survived vastly longer than man and is entrenched in every part of the earth from polar tundra to the deepest beds of the deepest tropical oceans.

The race, as elsewhere noted, includes some of the most ruthless destroyers of roots and the most harmful parasites of birds, fish, insects, and mammals, including people. The largest species thus far identified—sometimes as long as twenty-seven feet—infests the bodies of whales. Several of the smaller nematodes, no longer than one seventy-fifth of an inch, prefer people. No fewer than thirty-two species, including pinworms and hookworms, cause human diseases, including anemia and trichinosis.

There are also beneficial nematodes, tremendous numbers of them, including perhaps hundreds of species that prey directly on other nematode species that grievously damage plant roots and other life forms. If one goes along with what used to be the pathologist's credo that all soil lives are either good or bad, or even if one doesn't, there is evidence that the great majority of the known nematode species, almost certainly well above nine tenths of all now known, are beneficial or, at the very least, not injurious.

But here again the bad nematodes hog the limelight, and the principal damage tolls reported are those relating to agricultural crops, preponderantly the roots. As this is written, at least thirty valuable crops are known to be directly vulnerable to nematode damages, which in the United States are estimated as being as costly as a billion dollars a year. The branch of science which is now boldly but no doubt correctly called nematology has emerged within the present generation. Tobacco roots were among the first widely noted victims of the ubiquitous threadworm. Next, the commercial okra crop, tomatoes, and (well to the south) bananas required protective measures. Then the ever-important potato crop, which remains the most frequently eaten vegetable of the Western World, was tellingly stricken. One of the most widely reported onslaughts by nematodes was the harassment of West European potato crops by the rather large and well-named golden nematode.

This reached a climax during the 1920s. By the late 1930s much of Europe's best potato lands was nematode-fouled to a point where a potato crop could be planted only once in four or more years. Certainly this does not mean that nematode injury to potato crops dates only as far back as the first quarter of the present century. Most probably there have been potato-injuring nematodes as long as there have been potatoes, perhaps much longer. But ironically and all too expensively, the awareness of nematodes, and soil pathogens generally, is based on injuries done, rather than injuries foreseen or prevented.

In 1941 the golden nematode was reported for the first time in the United States—on two potato farms on Long Island. The symptoms were unusually conspicuous. Great areas of green vines turned sickly yellow and ceased growing. By 1946, New York State and the U. S. Department of Agriculture joined in an interesting effort to check the invasion. Soils were fumigated, seed potatoes sterilized, shipments were embargoed. By 1960, after the spending of about ten million dollars of public money, the Long Island potato crop was pronounced "saved"—at least from the golden nematode.

Meanwhile, there were harmful invasions of root nematodes. Beginning about 1950, the soybean cyst nematode began appearing in North Carolina. Two years later it emerged as a principal crop enemy in Arkansas, Kentucky, and Mississippi. Examination of the bean plants showed their roots infested with tiny worms about one twenty-fifth of an inch long, behaving in ways characteristic, though not invariable, among nematodes. Each nematode has in her mouth (most are female) a tiny needle, much like that of a hypodermic syringe, with which she slits open a root epidermis. Burying her head in the slit, she injects into the root tissue an enzymatic fluid to help her digest the plant juices, and so begins to feed.

As she feeds, her projecting rear swells and exudes a jelly-like substance in which she lays eggs, which soon hatch and join the layer in the onslaught on the roots. The cyst nematode is almost unbelievably prolific. She literally bears herself to death, producing egg masses inside her own body until it is a distorted mass. Then she dies. Her skin browns and hardens to form a cyst which attaches the eggs to the invaded tissues of the roots. The eggs within them lie dormant for years, until conditions are favorable for hatching. The development of chemical compounds strong enough to penetrate the cysts without killing the plant is the first challenge to manufacturers of nematode killers, but in rather impressive part the challenge is being met.

The symptoms of nematode attacks on roots are immensely varied and in many instances quite difficult to diagnose. Such evasive pronouncements as "spreading decline" are typical. In many instances the gradual running down is caused not so much by the root nematodes as by the fungi and bacteria that enter the roots through the nematode's puncture.

The "spreading decline" of Florida citrus orchards was first reported in 1926 in the area of Winter Haven, where only a few orange trees were visibly infected. When the sick trees were dug up, they showed rotting roots. While soil scientists and others were studying and striving to isolate the various fungi and bacteria present in the rotting, the infection spread inexorably to some 1300 groves in nineteen counties of Florida.

In 1953, plant scientists finally identified the culprit as the burrowing nematode, whose puncturing of the citrus roots opens the door for decay-causing organisms. Within the next five years, and with or without landowners' consent, Florida State Plant Board men equipped with bulldozers and power winches literally ripped out about 6500 acres of severely infested orange trees. Subsequently this drastic treatment, which causes many growers to vow that the treatment is worse than the disease, has been extended to several thousand more acres of citrus, including tangerine, grapefruit, and hybrids. After the trees are pulled up and burned, the land is deep-plowed, treated with a powerful fumigant, then given a rest or furlough of at least two years before being replanted with citrus.

As a branch (perhaps rootlet is a better designation) of zoology, nematology, however ancient its subject, is still an infant science which unavoidably does considerable mewling and puking in its mother's arms. As yet there are probably no more than 200 accredited entomologists in the entire world, and most of these are still listed as specializing zoologists or pathologists. With well over fifty known species of nematodes for every nematologist now available, the field of learning is huge, important, and appallingly undermanned. This holds true for many other entries in the sciences of soils and roots. There is progress to report. Research and manufacturing chemists are performing ably with materials at hand; so are farmers and orchardists.

The techniques of protecting crop roots from nematodes show especially promising ways ahead. In the main, they are based on building up the organic contents of the soil primarily with animal manure and decayed vegetation; then on strengthening the resistance of the root systems and attracting and encouraging other nematodes, fungi, and related soil lives, that resist or entrap the destructive nematodes.

Another defense is crop rotation—moving a crop to a new site before the population of destructive nematodes has time to build up. Spading, turnplowing, or otherwise twin-plowing the soil in hot, sunny weather also helps, because many of the root-puncturing species cannot withstand drying. Breeding nematode-resistant crop varieties, though tenuous from a standpoint of specific goals, likewise offers hope. To cite but one example, varieties of alfalfa with root resistance to nematode invasion are already in extensive planting.

Chemical control—use of soil fumigants that check populations of harmful nematodes without destroying the beneficial organisms in the soil—is thus far the most instantly effective defense. Antinematode compounds have been developed for direct use in the soil without serious injury to root structures.

The ever-tantalizing fringe land of well-known chemistry and little-known root chemistry is still another realm of brightening promise. This is the "root-scent approach." We are learning that the distinctive smells of certain roots definitely attracts certain species of nematodes and repel others. Many kinds of vegetable nematodes, for example, seem to abhor the root scents of asparagus or mustard. Many nematodes that would ravage corn roots hurry away from rutabaga roots. There is the distinct hope that the strategic planting of herbaceous lures and repellents may eventually be developed into an effective defense against nematodes and perhaps many other already proved or forthcoming destroyers of roots.

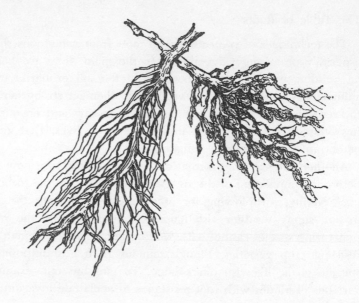

WANTED: HEALTHIER ROOTS

Sooner or later, usually sooner, a student of roots meets head-on with the grim, blunt truth that so far as anyone knows for sure there is no such thing as a completely healthy root structure. Undeniably, this is a prime reason for the broad but expert assertion that nobody has yet seen a completely healthy plant. Thus, instead of effervescing about hoping he might live to be a hundred, one might indicate his wish for a very long life by asserting that he would like to live long enough to see what truly healthy vegetation looks like.

"Healthy," of course, is a relative term; even in the animal kingdom, which is much less disease-ridden than the vegetable, "healthy" certainly does not denote or even imply an absolute optimum or 100 per cent status of health. The current outlook is that we may have *almost* healthy people long before we have a first look at *almost* healthy vegetation.

Dominant as they are and provident as they are, roots are unquestionably the least healthy portion of the less healthy kingdom of life, the vegetable. When they are pinned down and prodded for a percentage estimate of root health, few plant or soil pathologists are disposed to estimate the attained maximum as higher than 50 per cent of the theoretical and as yet unattained optimum. Some believe that in time and with improvements by man a level of root health of, say, 80 per cent of optimum may be attainable, but a prevalent estimate is that present levels of plant health are much closer to 40 per cent than 80 per cent of a reasonably well-disciplined vision of the optimum of vegetable health.

Dr. William Snyder, chairman of the soil pathology department of the University of California (Berkeley), which has both the largest and one of the most highly respected faculties of pathologists of all universities, makes the forthright point that because of eroding, thinning, and deplorably abused soils the over-all level of root health is not improving markedly, and that quite probably, as a general proposition, it is declining. Like many other scholars of soils and plant roots, he asserts that by means of superior conservation and upbuilding of soils, and with enlightened plant breeding and determinedly improved agronomy, levels of root health can be raised effectively. Even if the attained improvement of plant health is no more than 2 or 3 per cent or at the outside 4 per cent, of the maximal, it could result in doubling, or in some instances tripling or quadrupling the harvests from the field run of already proven crops. This educated guess is supported by revered and highly practical research now in progress in both hemispheres and many nations.

Thus far we have no real census of root diseases. Some highly respected pathologists have guessed the total of vegetative diseases as no fewer than 700,000. Regardless of the actual total, one can be reasonably certain that a whopping majority relate directly to roots. Some of the more competent directories of vegetative diseases have added better than 14,000 entries during a single recent year. Plant pathologists have long since run out of names. (A typical example is a recently entered tuber disease thus far known as "healthy potato virus disease.") The U. S. Department of Agriculture records now enumerate approximately 120,000 plant diseases, but make no claims of being anywhere near complete. Of the now duly recorded 2000 principal diseases of the thirty-one principal crops of the United States, about 91 per cent are definable as diseases of roots or root spheres.

To repeat, this listing of primary plant diseases is not presented as complete. It includes only those diseases currently being studied or already listed as consequential. Increase is not only expected but known to be occurring. Department of Agriculture scientists are definitely in position to confirm that more than nine tenths of plant diseases now being combated extensively relate to root structures and/or rhizospheres, and that this ratio is tending to increase.

Though admittedly we do not know enough about the pathological backgrounds of vegetative diseases, we are fairly well informed about where to look for root pathogens. There is no real doubt that a majority, indeed a preponderance, of the identified enemies of plant roots live in the upper six inches of the soil. By way of atonement, the same stratum, which is approximately the upper half of what soil scientists now call the A horizon, also contain the greater part of the nutrients and moisture on which plants live, as well as the greater proportion of bacteria and fungi known to be beneficial to root structures.

The undersoil or B horizon usually reaches several feet (three to four feet is average) below the A horizon. For most of our perennial crops, including the stronger grasses and many other forage plants, for orchard crops, for ornamental, timber, and forest trees, and for many principal field crops, the lower or B horizon is of profound importance, as a rule not so much in terms of organic or convertible nutrients, but because needed mineral salts and water are stored there.

Still lower is a third zone, the C horizon, which is likely to extend ten or more feet below the surface. Here soil is created for use in future eons. Its earlier designation, "nonsoil stratum," is therefore a double misnomer; future sources are indispensable components of soils, and we are quite certain that the roots of most of the large woody plants and many herbaceous ones probe deeply into this lowest horizon.

These deep-lying layers of soil are comparatively free of disease-causing organisms. One of the most esteemed of present-day microbiologists, Britain's S. D. Garrett of the faculty of Cambridge University, finds that about 60 per cent of all known soilborne diseases are initiated in the upper three inches of topsoil. Here, then, less than a finger's length from the surface, live most of the dominant pathogens of seeds and seedlings, including *Pythium* and *Rhizoctonia*, which cause various and ruinous root rots. Here also lurk most of the fungi that cause stems to rot, among them the wicked legions of *Fusaria*, *Sclerotinia*, and *Phytophthora*.

Dr. Garrett is convinced that about 30 per cent of all other root diseases thus far classified are unwelcome residents of the other levels of the A horizon—more than three inches but less than twelve inches below. Thus, about 90 per cent of all root enemies now know life in the first foot of soil. Destructive nematodes, or threadworms, are also found in the last-mentioned level, including those that cause ruinous galls on roots.

"The normal crop," says Dr. Garrett, "is one which suffers appreciable and regular degrees of root damage by parasites. . . . We can conclude that increased growth response means inevitably the better control of unknown or comparatively unknown plant inhibitors in the actual or more immediate rhizosphere."

In this, one of the most eminent American soil pathologists,

W. A. Kreutzer, now of the University of Colorado faculty,
concurs, and adds:

> It is probable that crop losses from low-grade and virtually
> unknown plant inhibitors may be many times greater than those
> recorded for the more spectacular plant pathogens. . . . In
> view of advances already under way, it is quite probable that
> soil treatment may eventually change our prevailing concept of
> plant health and disease. Perhaps someday we shall know just
> what a really healthy plant looks like.
>
> Most of the knowledge we have of roots is still based on our
> struggles to learn more about root diseases. I do not expect
> this circumstance to change, but at least we can keep on study-
> ing and striving with the reasonable hope that maybe our
> great-grandsons and great-granddaughters, if they are so dis-
> posed, will have some relatively healthy roots to study.
>
> The big job facing agronomic sciences today is that of help-
> ing root structures control or resist their enemies. The scope of
> study can no longer be limited or concentrated merely on the
> more spectacular fungal rots, rusts, and blights. The challenge
> now calls for searching out the more widespread causes of root
> diseases. We need to know a great deal more about soil bac-
> teria so as to know much more accurately which are harmful
> and which are beneficial; they are bound to be one or the
> other.
>
> As we advance along these lines, we will keep shifting
> toward preventive practices, keep centering our studies on
> smaller masses of soils, and paying more heed to the circula-
> tion system of roots. And we will keep moving farther away
> from violent treatment of soils with poisons. Instead, we shall
> try to help roots help themselves to better health.

Combating parasitic fungi in the soil is, of course, only one
of many approaches to the problems of root health, but it is
deserving of heed. As a rule and most fortunately, a given
parasitic fungus exploits only one or a very few species of
plants; usually its visible injuries are limited to a given portion

of a given plant species. The dour point here is that even if only one portion is visibly damaged, great harm is frequently done to the plant structure as a whole. Corn smut is a typical and all too convenient an example. The pathogen fungus grows and fruits in the tissues of the ears and the tassels where it produces large tumor-like swellings. It takes its nourishment from the stalk but perpetuates its evil kind by dropping its spores upon the soil. The spores germinate in the topsoil sphere, and from that vantage wait to launch another attack.

Most home gardeners are regretfully familiar with some of the readily identifiable root diseases and with at least some of the "spectaculars," such as clubroot rots which destroy beets, turnips, rutabagas, cabbages, cauliflower, and other home garden stand-bys. All these remind one who would correlate gardening and history that most of the more dramatic and far-sweeping plant diseases have thus far been caused by fungi.

But plant pathologists continue to reiterate that the less spectacular fungal pathogens, particularly those that strike at root structures, are of greater concern to growers and of greater peril to food supplies than those that blast leaves and stems. A cabbage smut, for example, is usually easier to combat than the powdery mildews that afflict grains, clovers, berries, and orchard crops, primarily because the lower-order pathogens infiltrate the vascular systems of the plants from secure hiding places in the roots.

In any case, there is no real doubt that the best hope for defending man's food supplies lies below the ground surface. An impressive point in evidence is the commercial banana crop, which has recently effected what is perhaps the most spectacular harvest increase ever recorded in a three-year period. Between 1963 and 1966, average banana yields in Honduras, Costa Rica, and Panama catapulted from about 350 bunches per acre per year to well above 700 bunches, for a calorie total upward of twenty million per acre per year, around ten times the present average food-crop yield. This is due partly to the recent development and extensive planting of a prolific banana variety called Valery. Significantly, the key factor of increase relates closely to greatly improved care and selection of banana rhizomes and roots, and particularly to prevalent advances in safeguarding them from injuries from the soil-infesting disease (Panama disease) caused by *Fusarium cubense*, from harmful nematodes, and from the bacterial infection called Moko disease. One particularly effective defense measure is the controlled thermal treatment of the "bits" or rhizome planting stock. The more effective minimum-temperature disinfectant is steam. In the beginning stages of its use an elderly tropical engineer, George W. Bump, salvaged an old-fashioned, long-abandoned steam locomotive for use as the applicator and demonstrated that treating underground planting stock with a twenty-minute steam bath fairly well destroys most root-infection factors. There are paralleling or similar progress reports in root treatment of other tropical crops, including cacao trees (the source of our cocoa and chocolate), vine peppers, and ginger, as well as the nursery stock of the African oil palm, now one of the principal providers of our salad oils and margarines.

In all instances the thermal sterilization techniques are supplemented with building up the organic contents of the soil. This serves to strengthen the over-all resistance of the root structures. It also serves to strengthen and encourage various soil inhabitants that help control the constructive invaders. Other improving defenses include crop rotation, the studied breeding of varieties markedly resistant to particular types of pests (such as the new nematode-resistant potato and alfalfa), and the scientific exploitation of root scents.

This is a passing reminder that we are beginning to acquire and effectively put to use the knowledge that the odors of certain roots repel many natural enemies of other roots. Many species of vegetable-harassing nematodes, for example, abhor the root scents of asparagus. A discreetly placed row or cluster of asparagus, or even mustard plants, may greatly aid nematode-afflicted okra or turnips. The odor of rutabaga roots is positively revolting to corn nematodes. The strategic planting of herbaceous lures and repellents is getting to be a more and more effective defense against underground enemies and ruiners of roots.

The more and more widely prevailing gist of promise in the vast, baffling area is that people are perceiving and applying more and more ways of safeguarding and improving root health, and more and better means for helping the all-helping roots help themselves.

8

BETTER BREEDING: BETTER ROOTS

John Weaver, the great American unearther of roots and ecology, was not the seventh son of a seventh son. He made no pretense of having extraordinary prophetic powers, not even campus clairvoyance. However, as a probing scientist the pick-swinging Nebraskan came to be a kind of walking bastion for the general belief that the hardiness of roots is a first denominator of sustaining harvests, beginning with our foods and fabrics.

Weaver was content with presenting painstaking supporting evidence. When one thoughtfully reviews his volumes of evidence and files of root photographs, one finds himself previewing a great many specific developments of now duly proved hallmarks of root productiveness.

Since 1920, John Weaver's most successful year in terms of

his publications, U.S. harvests of food crops have already increased by an average of 70 per cent, while the records of
other nations, including Australia, New Zealand, Israel, Denmark,
Holland, etc., are even more impressive. But to stay within
home boundaries, we can and perhaps should note that since
1920 the productiveness of a man-hour of farm work has grown
by about 500 per cent, and even during the 1960s U.S. farm
production, again on a basis of man-hour production, has out-
gained the average industrial production about two and a half
to one.

At least as far as I can discover, John Weaver did not fore-
cast such a present reality as the fact that by averages one
professional farmer in the United States now produces enough
food harvests to feed himself and thirty-nine other people. (This
is certainly no world record: one Australian or New Zealand
farmer now grows enough to feed himself and about sixty
others.) In any case, he perceptively recognized the all too
prevalent illnesses of roots, and Nebraska's distinguished root
scholar also recognized at least some of the superior possibilities
for root improvement, both natural and man-encouraged.

Certainly Weaver foresaw that better knowledge of roots
could redound to the improvement of virtually the entire range
of useful crops in fields, orchards, gardens, flower beds, lawns,
meadows, and pastures and on ranges; the last, of course, are
essential to the all-necessary development of meat and milk livestock. He was distinctly aware, too, that roots are the most
decisive areas for the genetic improvement of plants.

Not many life scientists in 1920 would have predicted that
within forty years the enormously virile young sciences of plant
breeding, very largely on the basis of roots, would be responsible
for the development of some 300 new commercial varieties of
grains, grasses, fruits, vegetables, flowers, or "ornamentals" every
year, or that fully a third of this ever-amazing global argosy
of new crops would be originated in the United States. But
John Weaver profoundly respected the potentials of plant

breeding, and he was quite certain that its progress would grow from hardier, more vigorous root structures of the entire range of economic plants from window-box nosegay providers to towering timber trees.

These were among the many items of what he termed earth writing that John Weaver deciphered while sweating out his root "profiles." Appropriately he lived to see a great new generation of plant breeders paying much greater heed to roots and the magnificent if still somewhat bewildering roles they play. He foresaw correctly that the future techniques and procedures of plant improvement would be centered on the living plant as an integrated unit of life. Even so, plant breeders are more and more agreed that the better-flavored apple, strawberry, grape, or onion, and the more beautiful rose or geranium, the more nourishing wheat variety that mills better and tastes better are the results of sturdier, healthier, more adequately rooted plants. All this shows acceptance of John Weaver's foregone conviction that root vigor, preferably inherent but nonetheless capable of improvement, is a basic goal in competent plant breeding. Attaining that goal involves efforts to increase root resistance—not just against a specific disease pathogen, but against the complex or multitudes of natural enemies that are ever present in soils.

Sugar beets illustrate particularly well the toils, triumphs, and unexpected discoveries that mark the quietly impressive progress in plant breeding—necessarily with prime emphasis on roots, since the sugar beet is now one of the most valuable and among the most widely grown of the root crops.

After Fidel Castro and his hard-driven yessers (or could it be Castro himself is the hardest driven of all of Cuba's slaves?) had succeeded in taking over the beautiful and tragic big sister of the Antilles, more factually describable as the sugar bowl for the world, alternative sugar sources leaped into almost worldwide demand.

Any student of sugar sources knows that the giant sweet

grass called sugar cane is far and away the most efficient source of both edible sugar and industrial molasses. As already noted, primarily because of improved root hardiness, sugar cane has long since changed from an annual or biennial crop to a comparatively vigorous perennial which, particularly in Cuba, keeps rising from the same roots for half a century or longer. Even the Castro gang men cannot frustrate the now almost immortal sugar-cane roots. But maintaining the practical status of an international sugar supplier requires other effectively managed facilities, such as deftly managed grinding mills, railroads, processing centers, etc., and the Castroites have not been able to deliver the necessary goods and services.

The next-best source, of course, is the sugar beet, whose revival necessarily shifts the sweets story from the deep-soiled tropics back to the temperate zones. The first so-called big-money markets for sugar originated in Europe, where in the main, sugar cane does not thrive. By the beginning of the eighteenth century several European countries were taking on the sugar beets as an important crop. In time, an undersized dictator named Napoleon Bonaparte took up his sword or club in behalf of the sweet beet that is of double value as livestock feed and as a sugar source. In due course Napoleon's Bavarian-schooled nephew Louis resumed the herbaceous proselytizing.

Though the sugar beet is technically a biennial, since it develops its seeds during its second year, practically speaking, it is an annual crop. By telling contrast, sugar cane continued to entrench as a tropical or subtropical perennial, powerfully sustained by superior roots. Early in the present century the progress of sugar cane was further augmented by brilliant genetic improvements effected in Holland's magnificent Proofstaaten Oost Java (East Java Agricultural Experiment Station), which presently succeeded in upbreeding root structures powerfully resistant to mosaics and other subsurface diseases, both fungal and bacterial. The sugar beet, meanwhile, continued to gain growth range, including widespread acceptance in the Great

Plains of the United States and Canada. But the natural, man-augmented root advantages of sugar cane continued to give the beet raisers and their political sponsors an extremely bad time.

Even so, after the Castro take-over, both government and private plant scholars in the United States, Canada, Holland, France, Belgium, West Germany, Sweden, Denmark, Finland, Soviet Russia, and other countries resumed efforts with plant genetics by making the sugar beets a more workable crop. The story is rather broadly significant of the entire greening, hope-raising field of genetic improvement of people-sustaining economic plants.

The most destructive enemies of the sugar beet remain the parasitic soil fungi and nematodes or threadworms (call them eelworms if you feel you really have to). Several nematode species determinedly pierce, suck, and breed in the big storage roots while the pathogenic fungi swarm in to further befoul the crop, in several instances following the lead of the deplorably aggressive nematodes—that is, the fungal pathogens take residence or nurture in nematode penetrations into the root skin.

The already developed protective measures such as chemical fumigation of the soil, the use of so-called trap crops to lure away the natural enemies, and the rotation or planned change-over to alternative crops are inadequate and quite typically so.

The sugar beet is principally a small-farm crop, and few growers can afford the equipment needed to apply soil fumigants competently. Nor can they afford to take long chances on trap crops, which frequently fail to lure the harmful nematodes effectively. Fewer still can afford the rotation procedure, which permits a beet crop to be harvested only at four-to-six-year intervals. Many farmers use the tops and mill refuse of the beets as dairy feed; this makes the situation even more acute, since poisonous spray materials cannot be used safely. Clearly, the solution is to develop varieties with inherent genetic resistance to the more damaging root enemies.

Progress is being made in improving the sugar beet, but some of the findings made in the multinationed beet-breeding effort may be even more important to world agriculture than to the crop itself. By 1960, ventures in sugar-beet improvement were under way in at least thirty experiment centers in nine countries. One of the first significant findings reached into the very genesis of plant breeding and root-improvement efforts—the seed.

At the Michigan Agricultural Experiment Station, a plant physiologist, F. W. Snyder, and two agricultural engineers, O. R. Kunze and E. E. Hale, offered detailed proof that the rough handling of sugar-beet seeds results in germination losses as high as 20 per cent and growth retardation up to 50 per cent of normal seed. The seeds are seldom damaged in the usual planting procedure. The injury occurs during the now routine processes of mechanical threshing, conveying, and storage. This finding strongly confirms the suspicion that much of the trouble that besets crop roots is due to avoidable injury of the seed embryos before planting.

Meanwhile, since 1965—especially in Holland, France, West Germany, Soviet Russia, and the United States—newly developed strains of sugar beet have been showing much greater resistance to both nematode and fungal pathogens. The soil fungi most harmful to the roots are mainly of the "root-rot" genus *Rhizoctonia,* while the nematodes are usually of the cyst-forming types.

In experiments begun in Salinas, California, agronomist Charles Price of the U. S. Department of Agriculture staff, and his helpers made selections from hundreds of thousands of sugar-beet plants, representing practically all of the varieties now in use. They soon discovered that the attacking nematodes and fungi almost invariably infest the same soil. Confirming reports came from at least twelve government and independent research teams abroad. And from fully as many sources new and hardier kinds of sugar beet are now emerging. Since the afflicting nematodes and fungi work together, resistance to either foe gives resistance to both.

This newest revival of sugar-beet research has thrown light on one of the most baffling mysteries of root life—the compatibility or incompatibility of the root systems of various crops when grown together or successively in the same soil. Largely by chance, people have learned that the roots of some crops thrive beside those of certain other crops, while still others somehow lose their stamina.

For one very widespread example, corn (maize) and soybean are among the enduring Damon-and-Pythias associations of cultivated plants. When interrowed, both crops usually thrive, their roots amicably sharing moisture and soil nutrients. Furthermore, the organic nitrogen provided by the leguminous roots of the soybean benefits the corn. All this demonstrates a sort of herbaceous internationalism. Corn is a native of the Americas —specifically, so many believe, the high plains of central Mexico; others surmise that it may have originated in the highlands of Guatemala. The soybean apparently originated in tropical or subtropical Asia. Only comparatively recently did the two crops come together. Yet in the United States, to cite just one example, the compatibility of the soybean to corn—this within our own century—has catapulted soya from somewhere near fortieth place to a secure third place among principal U.S. crops. It is axiomatic now to say that where one thrives the other will thrive, too. Beyond any reasonable doubt, this most benefiting compatibility of the foremost grain and the most provident source of vegetable protein is a phenomenon of root life. Its reach of significance can hardly be exaggerated.

The roots of most of our cherished garden vegetables associate happily, but as yet the same does not hold true for those of all our valuable crops. Some root structures simply cannot get along with others. This may seem whimsical, but there simply has to be a down-to-earth reason for such incompatibility. Accumulating evidence suggests that the apparent antipathy of certain root systems to others is based at least in part on the inadequacy of certain minor soil elements commonly known as trace minerals and of other elements not usually in generous supply.

Corn, mighty maize, our largest field crop, and the sugar beet, our number-one true root crop, have highly incompatible root systems. Observant farmers keep on complaining that when corn follows sugar beets in crop rotation, the corn crop almost invariably suffers. In several countries, again including the United States, plant scientists are searching for answers and remedies. Among the first to offer a firm explanation is L. C. Bowan, U. S. Department of Agriculture plant geneticist and physiologist who has done most of his research at the Washington State Experiment Station.

Bowan has proved quite convincingly that the failure of corn following sugar beets is due to a lack of zinc, one of the lesser plant nutrients. For reasons apparently not yet understood, sugar-beet roots cause chemical changes in soluble zinc that make it less available to other or subsequent plants. When deprived of sufficient zinc, corn plants show poor color and develop stunted stalks and leaves.

The precise chemistry seems to be intertwined with other chemistries of the still extensive twilight zones of knowledge of root lives. Zinc deficiency in soil has been blamed traditionally on an excess of phosphorus, but new experiments do not confirm that thesis; in fact, they openly deny it. Nor do they indicate that the sugar beet takes more zinc from the soil than do many other plants. Furthermore, the act of returning beet pulp and leaves to the soil fails to remedy the trouble. The appearances are that the root exudates of beets, by some reaction or chain of reactions, simply corner the zinc supply. Fortunately, this particular deficiency in the soil is easily corrected by applying zinc sulfate. This fertilizer (or better say, plant food compound) is readily available; and a comparatively tiny amount, about ten pounds to the acre, is usually sufficient for most crops.

But the growing scope of soil-nutrition studies continues to tell that in terms of food acceptance or demands, roots are certainly not easily placated. Plant geneticists agree that it is practically impossible to "breed away" a plant's need for any of the nutrients that its roots seek. Even so, the fact is now generally evident that breeding procedures can improve the efficiency of root structures in fulfilling their chemical needs. This attainment, however, prods us into wakefulness to the fact that for the benefit of plants and the animals they feed, including ourselves, we need to learn far more completely and exactly all the nutrient requirements of plant roots not overlooking the trace minerals. More and more of these keep showing up—not as something "new," but rather as members of the soil community that were present all the time but did not make a show of themselves.

Efforts to achieve higher levels of root health by plant breeding are directed along a number of markedly different lines. As an interested onlooker the writer has observed about two dozen experimental or research projects now in progress. One fairly typical enterprise is the distinguished experimentations of Ross C. Thompson, a plant pathologist and geneticist of the Agricultural Research Service of the U. S. Department of Agriculture in Beltsville, Maryland. For a third of a century Dr. Thompson has been, as he puts it, "working to revitalize old crops" by skillfully crossing or hybridizing them with hardier, better-rooted native plants of the same genus. He specializes in common garden vegetables, beginning with lettuce.

When centered in large blocks or "genetic areas," root research is most heavily concentrated on the grasses. There are excellent reasons for this. The grasses (*Gramineae*) are the largest and most widespread family of flowering plants on earth. Their importance simply cannot be overestimated; they are the great suppliers of our bread and meat and milk and eggs as well as the principal saviors and beautifiers of our hard-used earth. Among the more general goals of geneticists are more vigorous perennial forage grasses, greater insect resistance, larger yields of more viable seeds, and, eventually, more and longer-lived perennial grains that can materialize only as still more vigorous root systems. Specific goals include varieties with greater resistance to particular pathogens, notably soilborne fungi that cause rusts, smuts, and crown rots; improved winter hardiness; better tolerance of heat and drought; improved seedling vigor; a better capacity to compete with associated plant life; and better roots, sustaining lusher and faster-growing top structures and bringing about longer productive life.

John Weaver's studies of prairie grasses, with their almost immortal roots, did a great deal to stimulate awareness and enthusiasm for the steadily expanding greening world of grass. More and more plant scientists are seeing the earth at large more and more as the perceptive Nebraskan saw his own special part of it. Journalists like to speak of our time as the Space Age; more and more naturalists are inclined to regard it as the returning Grass Age.

Plant scientists are accepting the Weaver thesis that the roots of native grasses are a superior index to the soil and the climate of the areas where they grow. With wondrous precision grass roots adapt themselves to soil conditions as they find them; perennial grasses, as a great throng, demonstrate with exceptional pertinence the capacity of roots to accumulate food reserves for use in advancing and perpetuating leaf growth.

At least 400 genera and certainly no fewer than 6000 species or subspecies are now included in the grass family. Of these somewhere near 300 species, including many now in very extensive use, have been developed, adapted, or improved by genetic means during the present century. As a result, desirable forage and grain grasses have been transplanted successfully from country to country and continent to continent. Of the sixty grasses that are now grown most extensively in the United States, forty-three were brought here from other lands.

When the nation was young and limited to a comparatively narrow and principally forested strip along the Atlantic Coast, practically all the pasture and meadow grasses sown were from Western Europe; many were brought directly from the British Isles. Selective breeding of particular strains or varieties followed in logical procession, with people choosing with increasing care what seemed to them the superior planting materials. In the beginning the usual procedure was to collect the seeds of a chosen species or variety, plant them, and so establish a substantial collection of mother plants as a seed reservoir. Growth of these was watched carefully; those that were weak, deformed, or diseased were pulled out. Growth rates were compared, and response to rainfall, humidity, and temperature duly noted. The amateur breeders trusted a great deal to luck, but still more to the oftentimes superior ability of grass roots to adapt to different soils and climates. Failures were forgotten quickly and easily, while the eminent successes, such as timothy from Southern France, Kentucky bluegrass from Southern Germany, and Bermuda grass probably from India, helped keep the nation green, and sustained our grazing livestock and otherwise sustained and conserved our soils.

Grass breeding today involves the selection and multiplication of what are called maternal lines—strains with strongly inheritable characteristics, including big, strong roots. Routine seeding is now being supplemented—in some instances replaced—by clonal breeding. Clonal breeding consists of developing from a single chosen plant or group of plants successive progeny by means of root grafts, stem grafts, budding, or propagating by means of rhizomes, adventitious roots, or rooted culms. However, since the grasses are such a magnificent flowering family, the usual method of breeding is by pollination. Procedures most used include open pollination (without absolute restriction of pollen sources), top crosses (selections interplanted with a pre-established pollen source), and polycross (pollen sources restricted to selected plants with common and desired characteristics). The technical procedures become more and more exacting, and the exactitude is reflected clearly in the results

—better grasses, based on root systems of improved health, vigor, and resistance.

As the splendid work goes on, the goals of breeders are trending toward drought-resistant strains and others that are particularly well suited to the less fertile soils and to marginal or submarginal lands such as drylands, dunes, and desert edges. Another goal is grasses that grow well in the mixed stands that keep our lawns, parks, roadsides, and pastures green and attractive most or all of the year.

Tree roots are another area of notable progress but enormously difficult breeding work. For a long time genetic methods have been used very effectively for improving fruit trees. Most respond well to both bud grafting and root grafting, as well as to propagation by cuttings. By contrast, forest trees are often difficult to improve. Seeding techniques are not generally successful. Many produce seeds only at long intervals, and often the seeds are extremely erratic in terms of germination. Long-standard procedures such as bud or stem grafting are not successful in many valuable species of timber trees. In a great many species, unexpected variants, weaklings, and hard-to-explain throwbacks trouble and disappoint the would-be breeders.

Experts in tree genetics have tried without appreciable success to propagate various forest trees by means of cuttings treated with hormones that on other kinds of stems quickly produce roots. As one typical example, scientists of the U. S. Forest Service have attempted for years to develop root cuttings of yellow poplar (the tulip tree). Recently, in an experiment center in Oxford, Mississippi, foresters dutifully treated 1650 limb cuttings with the most highly recommended root-growth hormone. They did not get one "take" in the entire truckload.

Then a forester happened to remember how certain Great Plains grasses, which on occasion decline to produce seeds, are propagated by means of buds on roots or rhizomes. Since yellow poplars have neither rhizomes nor root buds, the forester tried the next-best thing: he removed and planted the sprouts that grew from the stumps of felled trees. The sprouts took root promptly, grew lustily, and were transplanted with excellent results. The still astonished experimenter cannot explain just why they grew. He can only point out that "what grows near roots" somehow produces roots. Within sixty days after transplanting, the stump sprouts grew about five feet above ground and five feet below. The value of this discovery could be tremendous. Simple as it may seem, it is unquestionably one of the century's better breaks in propagative forestry.

A different sort of problem is involved in current experiments that seek to keep fruit trees small, especially pear and apple trees. It has long been agreed that most orchard trees are now too big for the most efficient use. Not only are oversized trees difficult to harvest but they usually demonstrate that the proportion of flowering and fruit-bearing meristems, or areas of cell multiplication, tends to decrease as the size of the tree increases. Furthermore, during years of heavy yields the big tree's demand on its root structure is so tremendous that the next crop is likely to be very small or a complete dud.

Limiting the size of trees without limiting the size or sturdiness of the supporting root systems seems to many experts the best solution if it can be done. In experimental orchards, such as those of the Washington State Experiment Station at Wenatchee, workers are trying to keep young apple trees small and their roots of normal size. The men conducting this experiment —including a plant physiologist, G. C. Martin, and two horticulturists, L. P. Batjer and M. W. Williams—point out that they are not trying to instigate unnatural developments. They submit that modern horticultural practices tend to increase the growth rate and the size of most fruit trees and therefore make their yields erratic. Severe pruning rarely solves the problem.

The concept of dwarfed fruit trees is ancient. In the past, however, nurserymen have grafted stock on dwarfed roots to keep the trees smaller. Unfortunately, the dwarfed roots have sometimes turned out to be sickly roots, more than averagely prone to virus and other diseases even while in the nursery. Furthermore, their weaknesses have been passed along to the graft wood.

In the Wenatchee experiments, normal-sized and selectively healthy rootstocks are used. When the trees reach fruiting age they are sprayed with a hormone solution that retards the top growth without measurably retarding the growth of roots. (The retardant used is N-dimethyl amino succinamic acid, which is added to standard foliage sprays.) Apple, pear, and sweet-cherry trees and certain flowering shrubs are the subjects of these experiments, which by and large are showing good results and still better promise for the future.

The fairly immediate responses have been marked slowing of the top growth with little or no slowing of root growth, more flowers (anywhere from twice to twelve times the normal number), and fruit that is more dependably uniform in size. The new shoot growth is reduced by at least half. Successive harvests are more uniform in size, and the costly alternation of excessively heavy and excessively small yields is thus avoided.

The men who work with this experiment are becoming more and more certain that what can be termed the natural balance of important fruit trees is being re-established. They admit that neither the hormone compounds nor the application techniques now used can be regarded as finally proved, but the basic principle of restoring and improving root-and-top balance is being strongly confirmed.

Other experiments aimed at helping roots to better carry their loads of responsibility for the entire plant are concerned with the warding off of damaging insects and fungi. Some roots use chemical means to defend themselves quite effectively against insects. This establishes the general fact that the thrusts of insect depredation are directed toward the above-ground part of the plant, while the roots concentrate on sustaining the upper portion during the attack. This functioning, obviously, can be compared with that of a strong heart that sustains a man in a bout with pneumonia.

Nevertheless, gaps in the roots' subtle armor against insects are sometimes evident. An ominous example as this page is written (there is usually an ominous example while any page is being written) is a new corn rootworm. Reported first from Eastern Colorado in 1960, its area of infestation has spread at the rate of about fifty miles a year into the cornlands of Nebraska, Kansas, South Dakota, Iowa, Minnesota, and Missouri. By the mid-1960s at least fifteen million acres of corn were under attack by the rootworm, which has proved strongly resistant to chemical defenses, including the chlorinated hydrocarbons. More recently the attacking force has dwindled as better soil disinfectants and stronger-rooted hybrid corn strains have been developed and proved. The enemy insect follows an aggressive cycle. The rootworms breed in the topsoil and emerge as beetles in late July or August to feed on the corn silk. During the fall they deposit their eggs (as many as 300 per beetle) in the top six inches of soil. When the eggs hatch the following spring, the larvae implant themselves in the roots of the young corn plants. Throughout the growing season they feed on the roots, retarding the plants' growth and reducing or completely destroying the yield.

The first progress in developing a remedy relied on extensive use of a favorite chemical entree of roots in general—organic phosphates. The strategy was to increase the vigor and resistance of the total plant. The prevailing progress in defending our number-one field crop from a particularly alarming enemy is one of well-precedented genetic techniques from the selective breeding of resistant varieties with still hardier roots.

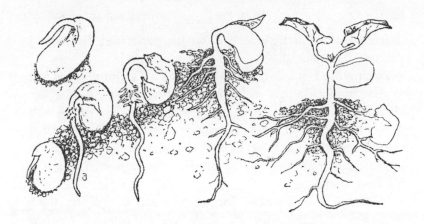

ROOTS AND LIGHT

For many centuries roots have been pictured as living in absolute darkness, hidden from sight, and therefore wonderfully mysterious. There is at least some mystic justification for this concept. There is also the growth of scientific correction. By human standards, roots usually dwell in darkness—at the very least, in comparative darkness. But we are learning that they are profoundly influenced by light and that soil, or even subsoil, however rife with mysteries, is not a phenomenon of total darkness.

Sunlight, whether it reaches the plant roots directly or indirectly, is their prime source of power. We are discovering with real certainty that sunlight coordinates roots with photosynthesis, motivates their growth, plays an important role in their metabolism, and enables them to carry on their unending work of collecting water and nutrients from the earth and distributing them to the plant above.

Tracing the beginning and growth of man's awareness of the relation of roots to light makes a fascinating game. Aristotle and his followers were among the many who played the game—with zest, one gathers. Down successive centuries others puzzled over the matter until bit by bit the picture began to come together and grow validly recognizable, although even now it is far from complete.

In the ever-lengthening line of perceptive students there were several British scientists of the early nineteenth century. One was Sir Humphry Davy, author of a great pioneering book, *On Agricultural Chemistry*. Almost simultaneously, from Queen's College in Cork, Ireland, there came Baron Justus von Liebig's memorable *Natural Laws of Husbandry*. Sir Humphry's prophetic volume included the then radical thesis that root systems are truly chemical laboratories, powered and activated by sunlight. Baron Liebig concurred and then settled down to the task of trying to explain it:

> As the food of plants cannot exist for any length of time in solution in soils, it is clear that there cannot be a circulation of such solutions toward the roots, but the latter must go in search of food, and that on virtue of power imbued from the sun. Hence the great importance of better studying the ramifications of the roots of plants cultivated by man . . .
>
> Plants have been justly described in general terms as animals turned inside out, the leaves representing the expanded lungs and the roots with all their ramifications, the absorbents of the intestines.

The Reverend M. J. Berkeley, at the time also of Queen's College, set out to record what he saw as a "great new movement in the understanding of natural history." In a pamphlet entitled "The Development and Action of the Roots of Agricul-

tural Plants at Various Stages of Their Growth," he pondered
the unique power of roots for effecting vegetative growth, and
ended with a discourse on the distinctive virtues of roots as
earthly saviors of man. Since "earth is redundant with poisons,"
Berkeley pondered the cleansing power of sunlight as one of the
special gifts of God. He wrote:

> It is well known that both vegetable and mineral poisons
> abound in every measure of soil, same being the self-evident
> floor of life. How, then, do the poisons act? . . . Roots do not
> absorb indifferently all substances which are soluble in water,
> or which may be incorporated in the soil in a minute state of
> division.
>
> An experiment may be easily tried with arsenious acid, and
> thereby the more reason to repeat the experiment, as published
> accounts of it are still very conflicting. The truth seems to be
> that if the dose of poison is only small, it is not absorbed; if
> large, the spongelets of the roots are destroyed, and a portion
> enters, but only a short ways up.

The Reverend Mr. Berkeley's appraisal of the ability of roots
to select, shield, and otherwise protect animal life was certainly
thought-provoking, but it side-stepped the capacity of root structures to utilize light or heat in their own specific terms. One
of the next cogent landmarks was the publication of A. Anderson's observations on turnips in the *Journal of Agriculture and
Transactions of the Highland Society* (Edinburgh, 1863). A
member of the local medical faculty, Dr. Anderson was especially interested in the capacity of roots to store nutrients that
they somehow gather from the soil. From his studies of turnips
the studious physician and teacher reported:

> In the first half of the time of vegetation the organic labour
> in the turnip plant is principally directed to the production and
> development of the external organs; during a second period of
> 35 days, 9 parts of the soil food absorbed went to the leaves as

against 2 parts to the roots; while in a third period of 20 days, 9 parts remained in the leaves and 16 parts [went] in the roots. During a fourth stage, the weight of leaves kept constantly decreasing, while that of the roots increased, the proportion being much larger than in the third stage. . . . Sunlight, whether received directly or somehow stored in leaf or stalk, is the propellant.

More decades were to pass before further contributions were made to man's understanding of the role of light in the growth of plants.

The final third of the nineteenth century brought forth several notable statements of appreciative awareness of root power, but with the controversial emergence of the works of Charles Darwin, the forces of study tended to shift away from the actual physiology or specific function of roots to the general sphere of roots. Darwin's imaginative defense of earthworms, as recounted in his rather fervent *Formation of Vegetable Mould through the Action of Earthworms,* included his pronouncement that earthworms each year bring to the surface a volume of "animal discharge" (worm "casts" or feces), enough "soil" to form a layer two tenths of an inch in depth or ten tons per acre. This hypothesis presently found its way into widespread public acceptance. The completely untrustworthy deduction was that root health could be practically preassured simply by leaving all to the earthworms. The engaging, bearded prophet from Shrewsbury had no qualms about the insidious talents of worms for spreading many kinds of pathogens of root diseases, and the same held true for his enthusiasms for the subsoil "vegetable mould." Darwin also blurbed the value of ants and the despised "ground rodents." But whatever else the Darwin hypotheses produced, nobody can deny they were helpful to the realization that soil is indeed a living institution and that most of its community is composed of beneficial lives.

More decades were to pass before the coming of concerted attempts to identify the power source and power utilization that must explain the ever-wonderful growth and function of roots. Then, principally from Germany and Holland, came important studies of root anatomy. One memorable revelation was F. Schwartz's "intimate examination" of root hairs. The Schwartz finding was that these single-cell tubes are superb instruments for absorbing fluids directly from the soil. He also noted quite specifically that the length of a root hair is rarely more than 1/150th of an inch, and that by averages, between 300 and 400 occur per square inch of surface or epidermal area. The German plant scholar painstakingly observed and recorded the amazing talent of root hairs for absorbing fluids. He further showed that where root hairs are present, a given surface of root can absorb twelve to eighteen times as much fluid as a root area without hairs.

In his prolonged and generally brilliant study of water in its relation to root function, Schwartz and his colleagues proved that many species of plants use from 200 to 800 pounds of soil moisture to produce one pound of organic matter above ground. The investigators began to wonder how such amounts were hoisted up to the tops of plants. What was the source of energy for this tremendous water lifting? The students readily showed that the rate of absorption increased with a rising temperature. Absorption of the sun's heat by the soil must then be an important factor.

As the twentieth century began, plant scientists of many nations were purposefully placing thermometers in various kinds of soil and seeking to make correlation between root growth and soil temperature. Some of the earlier trials are still highly revealing. The first conclusion reached was that as a general rule a topsoil temperature of sixty-five to seventy degrees Fahrenheit was the most favorable for total plant growth. When a visiting Dutch botanist asked Oxford's famed naturalist Paul Chalky how he explained it, Chalky roared: "A pure

scientist isn't supposed to explain, only to compute. So I say take
your bloody Centigrade soil readings, multiply by nine, divide
by five, list them in your bloody record book as 'Fahrenheit'
and shut up!"

It was not that simple. The Dutchman (Karl von Steuben)
verified the fact that roots in deeper soil grow vigorously at
temperatures much lower than sixty-five degrees Fahrenheit;
indeed, that roots of staple crops grow throughout a temperature
range of forty to one hundred and fifteen degrees, and that
deep perennial roots can still grow vigorously at temperatures
considerably below forty-five degrees F.

Meanwhile, in the United States, a resolute young life-scien-
tist and root scholar extraordinary, John Ernest Weaver of
Nebraska, whom we have already noticed, had begun a long
study of root temperatures and growth responses to heat. He
began with recordings of germination tests, such as the following:

TEMPERATURES APPARENTLY REQUIRED FOR
EFFECTIVE GERMINATION OF SEEDS*

Species	Minimum	Optimum	Maximum
Wheat	40 F.	84 F.	108 F.
Maize (corn)	49	93	115
Pumpkin	52	93	115

Having established the temperature range for beginning
shoots and roots, Weaver began to record the temperatures of
typical soils at various root depths. He set up observation
stations in Lincoln, Nebraska; Phillipsburg, Kansas, and Bur-
lington, Colorado. He chose recording dates to include four
progressive periods of root growth of common field crops: April
28–30, May 19–21, June 9–11, and June 28–30. The soil depths
at which he recorded temperatures ranged from half an inch to

* Weaver, J. E., *Root Development of Field Crops*. New York: McGraw-
Hill, 1924

four feet. Even his first findings confirmed quick-changing temperature variations in the surface soils and relatively stable
ranges at depths of more than a foot. The following are typical
records:

SOIL TEMPERATURES IN DEGREES FAHRENHEIT*

Depth in feet	April 28	April 29	April 30
0 to .5	59.7 F.	75.2 F.	53.6 F.
.5 to 1	55.4	73.5	51.8
between 1 & 2	53.6	59.0	50.9
" 2 & 3	53.6	57.2	50.9
" 3 & 4	51.4	53.6	50.0

	May 19	May 20	May 21
0 to .5	69.8	68.0	72.0
.5 to 1	66.2	64.0	59.0
between 1 & 2	56.1	61.5	57.2
" 2 & 3	55.6	59.0	54.3
" 3 & 4	53.6	56.3	53.2

	June 9	June 10	June 11
0 to .5	72.7	70.5	67.6
.5 to 1	71.2	68.4	65.8
between 1 & 2	68.4	65.8	62.2
" 2 & 3	65.5	63.0	60.4
" 3 & 4	64.0	60.8	59.0

	June 28	June 29	June 30
0 to .5	70.2	73.8	84.6
.5 to 1	70.0	73.3	79.0
between 1 & 2	68.2	70.7	75.0
" 2 & 3	65.8	70.2	71.2
" 3 & 4	63.3	68.9	68.0

The prowess of root spheres for holding and moderating heat
from the sun grew more evident as he continued the soil-tem-

* Ibid.

perature records. Seasonal averages maintained for ten consecutive years in Lincoln, Nebraska, were as follows:

Season	Air Temperatures at Soil Level	Soil Temperatures at					
		1 in.	3 in.	6 in.	12 in.	24 in.	36 in.
Winter	25.9 F.	28.2 F.	28.8 F.	29.5 F.	32.2 F.	36.3 F.	39.0 F.
Spring	49.9	54.8	53.6	51.7	48.5	45.7	44.3
Summer	73.8	82.0	80.9	79.1	73.8	69.0	66.2
Autumn	53.9	56.4	57.6	57.1	57.2	59.3	60.3

Continuing studies of root activity in relation to soil temperatures have indicated a number of diverse but generally pertinent truths. For example, as Weaver pointed out, back in the foggy 1920s:

Soil temperatures may have an influence on the presence of soilborne diseases and their increase. Pathogens are frequently unable to adapt to the higher or lower temperatures that suffice for root growth.

Soil temperatures are prime factors in conserving the supply of water in the soil. In general, when tillage is reduced and plenty of organic matter is included in the soil, the temperature will be lower and the water stored in the soil will be most economically used.

The soil-temperature range in a given soil tends to keep within the pattern known, or indicated, to have existed for a century or longer. The preferences of economic crops also tend to keep within well-established patterns and averages.

The common sugar beet (*Beta vulgaris*), which we have also mentioned at some length, provides an above-average demonstration of how crop roots adapt themselves to soil-temperature patterns. As also noted, the plant is a biennial, with the normal first year's growth concentrated on storing food in a big, fleshy taproot. During the second year, if the beet is permitted to grow, this storage function is completed and the seed crop is produced.

From the moment the seed of the sugar beet germinates, temperature, water supply, aeration of the soil, and available plant nutrients join conspicuously in deciding the growth rate. Within two months after the seed is planted, with medium growth conditions, a moderately healthy plant will have eight to ten leaves, a strong taproot (about a foot long in moist soil, nearly two feet in dry), and two well-defined rows of lateral roots.

By the beginning of the third month, the area of maximum growth is the taproot, which takes shape as a storage root which pushes downward somewhat like an oversized carrot or parsnip, occasionally forking. Lateral roots next develop directly beneath the soil surface, crowded closely together on all sides of the taproot and extending horizontally anywhere from six to eighteen inches. Being profusely branched, the side roots form a superb water-absorbing system in the upper foot or so of soil.

As the plant grows, additional and longer lateral or feeder roots continue to emerge from the taproot and to penetrate deeply into the subsoil. This complex of roots equips the plant to absorb water and nutrients and to make use of favorable temperatures at all levels from the surface to the maximum depth of root penetration. Certainly, the water supply obtainable is a decisive factor in shaping the root. In a dry soil the taproot develops a narrower structure and follows a more twisting course of growth. If the drought continues, the downward growth is extended. If rain comes, the roots grow even more lustily but with a much broader horizontal spread. In this area, sun power appears to be taken for granted; water supply sets the general shape and pattern of the root system as a whole.

However, and quite measurably, root growth definitely responds to prevailing soil temperatures. The sunlight that has warmed the earth also serves the plant in other ways. We continue to list photosynthesis; the creation of chemical compounds, particularly of carbohydrates, in the chlorophyll tissues exposed to light, a leaf function powered by sunlight, is wonderfully correlated with the activities of the root system. The light-on-chlorophyll compounding of carbohydrates from carbon dioxide and water profoundly affects root development even while supplying the roots with important nutrients.

Thus, roots become a principal participant in the light cycles of photoperiods that are so profoundly important to plant life. Measured in terms of photology, or use of light, plants can be classified as short- or long-day plants, or intermediate. A book—perhaps a shelf of books—could be written on this subject. We know that by selective breeding most useful plant species can be adapted to locally prevailing intervals of night and day. A typical vegetable garden today is an assembly of plants that have been adapted from a broad range of latitudes where photoperiods vary widely. Some of the preferred vegetables are from the equator, others from far-northern latitudes, lands well above forty degrees north or south latitude.

Their adaptation has been accomplished mainly by natural means and has therefore required centuries of time. Even so, cold-country plants have been rooted successfully in the tropics and tropical plants have moved triumphantly into the temperate zones and even into subarctic regions. Rather recently, plant geneticists have achieved brilliant successes in tailoring many crop and ornamental-plant species to the sunlight-and-darkness sequences of particular localities. In several instances, long- and short-day varieties of economic plants have been developed within individual species.

An exceptionally valuable example is the soybean. This great legume came originally from tropical borderlands of the Orient. During the twentieth century it has emerged as an important, widely planted temperate-zone crop. The prime chore of plant men working with the soybean has been that of developing light-adjusted varieties—more than a hundred in all—for locations with differing intervals of sunlight and darkness.

To look more closely by picking a typical example here in

the United States, the Biloxi soybean, a short-day variety, thrives
and bears splendidly along the Gulf of Mexico and in much of
the South. But in the latitude of Washington, D.C., and farther
north, it flowers so late that frost almost invariably destroys the
harvest. However, the continuing selection and upbreeding of
soybean strains having tolerance for varying amounts of sunlight
have established the soybean as the number-three field crop of
the United States and one of the world's first ten.

Basically, however, the soybean remains a short-day plant;
the development of its flowers require less than twelve hours of
light in each twenty-four-hour period for the development of
flowers. Unquestionably, the root shares this requirement. Long-
day species, such as most garden flowers of the north temperate
zone, fail to blossom unless they receive twelve or sixteen hours
of light each day while the flowering buds are developing. Recent research places the emphasis on the relative duration of
darkness rather than of light.

Plant scientists are still not sure how it happens, but are inclined to agree that the photoperiods preferred by a given plant
create a hormone in the leaves that is translocated to the ter-
minal buds (and perhaps to other growth areas), where it in-
duces the formation of flowers.

Flowering is, of course, a crucial phase of plant life, with
roots playing a part in the story, though here many of the details are not yet known. There seems, however, to be reason to
infer that both the temperature and the aeration of the soil are
important contributing factors.

There is reason to believe, too, that the soil may be substantially better lighted than is generally assumed. Smatterings of
evidence encourage this view; other evidence qualifies it. We
know, for example, that in a short-day plant such as the Biloxi
soybean the temperatures that prevail during the dark periods
have a more marked influence on flowering than those that
prevail during sunlight periods.

Night temperatures are highly important to root life, but the light-and-dark sequences may be even more important. The artificial lighting and shading of vegetation continues to gain importance, and these man-made variations of light and dark periods may open the way to greater abundance.

The comparative darkness of soil is related to still another life factor that is unquestionably influenced by both light and roots. This is the phenomenon called dormancy. In some ways, the condition is like sleep. Dormancy shows that a plant's urge to grow yields periodically—sometimes for long intervals—to an urge *not* to grow. A great deal is still to be learned about these urges to grow or not grow; unquestionably, they have a lot to do with deciding both the quantity and the quality of harvests.

Though roots are less subject to dormancy than other parts of plants, they seem to be the deciding factor in the motivation of dormancy. Leaf shed, the most conspicuous indication of impending dormancy, occurs in plants of all latitudes—from tropics to tundra. It takes place when the roots send a special substance up to the base of each leafstalk. A weak wall, called an abscission layer, is then formed, and when it is complete, the leaf is ready to drop. But the supreme masterpiece of dormancy is the living seed.

As man seeks ways to increase the abundance of his crops, he studies the seed carefully and plots to reduce the normal span of its resting period. Impatient about future supplies of food for his multiplying dependents, he protests what he considers the excessive dormancy of certain seeds and other planting materials. As any home gardener soon learns, some seeds simply refuse to sprout, no matter how carefully they are planted. Sometimes the seed coat cannot be softened by water, or the coating may be so tough that the little plant cannot break through however it swells. Other seeds have coverings that oxygen cannot penetrate—at least within a time that man regards as "reasonable." Sometimes high heat or freezing is needed to open seeds, and some seeds simply cannot germinate until their embryo plants have spent their allotted time inside them.

Granting that dormancy is essential to plants and their seeds, man is learning ways of diminishing the natural time lag it involves. Techniques used to persuade seeds to sprout sooner include such varying procedures as chemical treatment, direct surgery, soaking in water or water-soluble compounds, heating, storing at certain temperatures, and the selection and development of strains that are more agreeable to punctual germination. Obviously, all of these methods should be considered temporary and mainly superficial expediencies until we know more about them.

Plant scientists and others are making some progress in their attempts to solve the riddle of dormancy. An important step is that students of plants have confirmed the belief that roots play an important part in establishing the dormancy intervals as a fundamental of vegetative life, at least for the ever-dominant perennial.

Particularly in the temperate zones, the buds of woody plants do not usually expand in the same growing season during which they are formed. Those that began life during spring or summer almost invariably remain dormant until well into the following spring or summer. As a rule, roots do not go dormant but continue to prescribe for or influence the other parts of the plant.

Even those who qualify as competent chemists are still a long ways, though not necessarily light-years, away from understanding the creative competence of plant roots, but they are learning bits and pieces about it. We now know that by exposing stems or buds to such chemical compounds as ethylene dichloride or ethylene chlorohydrin we can modify the dormancy that these master chemists, the roots, have decreed. We are finding chemical ways to reduce conditions of dormancy that are established directly within root spheres—as with tubers (the common potato is sealed off from growth for many weeks), rhizomes, corms, and bulbs.

Generally speaking, however, the most effective and economical way to reduce the dormancy periods of seeds and planting stocks is to imitate the increasingly well-known methods of roots and rhizospheres. All this is valid evidence that the related magics of sunlight, roots, genes, and vegetative chromosomes and hormones continue to take form and utility as keys to far more generous harvests.

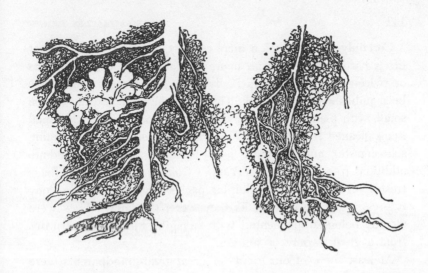

10

ROOTS AND CHEMISTRY

Beyond any real doubt, what people still don't know about roots could fill a major public library. As of this moment, what people really know about roots could almost certainly be stashed in one living-room bookcase, though there could be some reasonable question about the precise size of the living-room bookcase—that is, whether it should be the neat economy size that costs three and four fifths books of Green Stamps or the big classy "divider" that calls for nineteen books. There could also be some reasonable argument about the somewhat loose-lugged phrase "what people really know."

You and I have every right to our own estimates, but nobody can deny that we are learning or can learn more with every passing day and hour, and that relatively speaking, we are very markedly warming to the pursuit.

Certainly, to look back a mere century, which in terms of root life is hardly more than an hour, all that our great-grandfathers or twice-great-grandfathers really knew about roots could have been published in one book. That lone volume could have been small, with plenty of blank flyleaves to accommodate additional items gleaned from the *Old Farmer's Almanac*, the bright young schoolmaster, and the more lucid bartender, with intermediate additions from the pages of *Youth's Companion* or the mailbags freshly lifted off the Philadelphia packet boat or from the Pony Express pouch. For scant as they were, the known facts were already being supplemented with engaging theories or surmises from a great variety of sources.

At least some of our great- or great-great-grandparents were generally aware that a root system is a real-life chemistry laboratory, subtle, determined, virtually continuous in work, and very basically self-sufficient. Noah Webster, the great schoolmaster-lexicographer, had long since approved and helped popularize the phrase "living chemistry" and driven home the acceptance that chemistry is in fact the science that treats of both the composition of matter and the transformations that matter undergoes.

That expansion of context was and is most appropriate in terms of the wonderful, interminable saga of roots. The formal chemists of the past century found the feat of formulating what more recently became known as soil chemistry very difficult to accomplish and even more difficult to correlate with what had been labeled "pure chemistry."

But now a vastly strengthened new generation of soil chemists is earning accretion with biology, pathology, entomology, physiology, and practically all the other life sciences. In particular, for the cause and needs of learning about roots, chemistry and plant physiology are already in the throes of a kind of shotgun wedding. Their combined situation is like that of my eighty-three-year-old Vermont neighbor who recently found himself yet again headed churchward to be wedded—not because he chose to, but because, b'God, he was obliged to.

Fortunately we now have many gifted chemists who are entirely willing to enter into working wedlock with any other devoted scientists, including biologists. Princeton's renowned biologist John Tyler Bonner heartily returns the compliments by openly stating that he finds chemists practically indispensable to his progress, and obtains very special delight in what he terms "the clean satisfaction of finding the chemical structure of any compound which does important things within an organism." Dr. Bonner further reflects on why first-rate organic chemists are willing to devote so much of their time to helping biologists or physiologists with momentous problems, such as trying to comprehend roots. "It is an almost everyday occurrence," he adds, "for a biologist to discover that some plant produces a chemical substance which is of key importance to its own life's history . . . such as a hormone which has been isolated and upon injection produces some striking specific reaction."

The Princeton faculty man points out that what usually happens is that the biologist supplies the crude extract and the chemist proceeds to make an analysis, usually an extremely laborious and complex procedure, for the purpose of supplying the biologist with a structural formula. It is frequently a very difficult chore, and far beyond and above the chemist's assigned duties. But to mention only one group of immensely beneficial results, such voluntary comradeships have had most to do with producing such revealing attainments as synthetic plant hormones. These serve doubly the improvement of many valuable crops and the over-all progress in understanding roots.

Various other sciences are also helping most valuably to pry secrets from and about roots. One of the many brilliant entries is the microscopy developed from the magnificent electron microscope, which is helping mightily in the entire field of root scholarship; another is microclimatology; another, soil-level thermodynamics. These are merely a passing note in regard to the rapidly increasing number of approaches to the study of roots. But chemistry is the ever-accepting and -giving ally.

Root-directed soil chemistry began finding its way into textbooks back in the 1860s, with efforts to defend certain crops against the ruinous forays or total destruction by vegetative contagions, principally caused by fungal pathogens and in great part directed at roots. In the first and usually frantic defensive efforts, chemical compounds were scattered, dripped, or squirted into the soil without any particular consideration of or distinction between friend and foe among the multitudes of lives that compose the soil. Most of the procedures were lifted quite directly from prevailing practices in human medicine and sanitation. This held particularly true for the real or would-be fumigants.

Readers who, like this writer, are long past thirty-nine may remember the rather horrible practice of fumigating the sickroom, sometimes the entire house, after a visitation of any contagious disease. Carbon disulfide was the most widely used fumigant. We may never know how effective it was as a germ killer, but few will ever forget how it stank.

This same horrendous carbon disulfide was the first widely used soil fumigant. Its initial target was a devastating root pest of grapes, an infection carried by the plant-louse *Phylloxera*. During the 1860s the grape ruiner appeared in vineyards of France, then in Germany and Italy. The Thenard-Monestier method, which employed carbon disulfide, was used in an effort to save the withering vines. In some instances it was promptly successful. Dying vines began to leaf and bloom again, and some of the less severely afflicted showed restored vigor. But, alas, the so-called miracle cure was of brief duration. In any case, the temporarily beneficial effects confirmed the benefits from various sulfur compounds as protectors of roots.

By the 1870s, soil disinfectants were being used quite extensively in the battles against root pests and diseases. From remote Australia came the French method. It employed powdered sulfur as a cure-all. Nobody could deny that the originator, C. A. French, was at least a benefactor to the sulfur industry. In England, the Thaxter process also used sulfur as an aid to commercial gardening, as a repellent for root diseases or "wilts." Americans varied the sequence in 1900 when the Selby method was adopted in an attempt to save the New York State onion crop by drenching the planting rows with a solution of formaldehyde and water.

During 1913, a German pathologist, A. Reins, introduced a method of treating wheat seeds with a chlorophenol mercury compound to control various root pathogens, including the ruinous infestations of grain bunt or black smut. This pioneer venture in disinfecting seeds involved treatment of the small zone of soil surrounding them.

The next major advance in the chemical defense of roots came as a very direct result of the gas warfare practiced during World War I. The chloropicrins were the principal gases used in this particularly ghoulish type of mass slaughter, and chemists noted that the airborne gases, at least in some instances, were almost as lethal to fungi and insects as to man.

One of the early legitimate uses of chloropicrin gas was for the control of pineapple pests. By the early 1920s, cultivated pineapples in Hawaii were more or less chronically sick, from the roots up. The low-fertility soils that produce pineapples of the best size for canning were beset by fungi, nematodes, and other root pests that seemed to share an urge to annihilate, or at the very least decimate, the already sick crop. G. H. Godfrey showed that when the ground surface is sealed off with paper or some other effective covering, chloropicrin can be released fairly effectively in porous soil to kill many types of fungi and some species of nematodes.

The Godfrey findings prepared the way for the use of other volatile soil fumigants. Beginning about 1935, the lethal compound methyl bromide, which had already proved effective as an insecticide, was put to use as a nematocide—a killer of several species of nematodes, including those causing knots or lumps on roots. Chemists C. W. McBeth and A. L. Taylor, then of the U. S. Department of Agriculture (Dr. McBeth is now a senior soil chemist with the Shell Research Laboratories in Modesto, California), were the pioneers in the use of methyl bromide as an effective defender of roots. They employed the Godfrey sealing-off method for making the pesticide serve as an effective soil fumigant—prior to the formation of root systems. However, the cost was so great that only a few high-value crops such as tobacco and winter vegetables could justify its use.

In 1943, William Carter of the New Jersey Experiment Station identified and demonstrated a fairly inexpensive petroleum product that is also effective in protecting roots against nematodes and certain other animal enemies. This is the now common liquid petroleum gas, propane; it can also be used as a fairly manageable soil fumigant. In 1945, Dr. J. R. Christie, of the University of Florida and its various plant experiment stations, and now incidentally one of the first accredited American authorities on nematodes, identified and proved the effectiveness of another fairly inexpensive chemical killer, ethylene dibromide. When this volatile material was accurately compounded and put in use, the defense of roots made another notable advance chemically, and nematology, still one of the newer root sciences, had a fairly effective working tool.

The relatively little-heeded saga of better defenses for root health keeps moving ahead. The late 1940s saw a much more varied list of root-protective chemical compounds being made publicly available. Dr. Christie developed allyl bromide as another root defender. More recently many gifted petroleum chemists have come forward with a greatly expanded and steadily improving list of soil fumigants derived from petroleum. Granting, as one must, that the development of chemicals for soil treatment is still in its early stages—even, as some insist, its infancy—there is already valid proof that volatile materials that can be transported as liquids and applied as soil-penetrating gases are the most promising defenders of living roots as soil fumigants. These are now being supplemented by other quick-acting chemicals designed to reduce the ranks of a particular pest in the least possible time, and with minimum disturbance of adjacent soil lives. These are proving themselves the most advantageous of root-protecting strategies.

The list of these readily available compounds is growing rapidly; it includes less expensive antibiotics derived in most instances from fungi, organic phosphorus compounds, and various sulfur carriers and metallic salts. At this time the accent is on the development of root-sphere fumigants that are not soluble in water and therefore not vulnerable to rain, yet are capable of rapidly reducing the numbers of enemy fungi, insects, nematodes, or bacteria that are known to afflict roots and/or their rhizospheres. Many of the compounds can cope with two or more root enemies simultaneously.

Even so, the actions and reactions of soil fumigants are often puzzling and unpredictable unless supported by extremely competent tests. Methyl bromide, for instance, is highly destructive of weed seeds but will not kill fungi. The chloropicrins penetrate loose soil effectively, but not the tissues of many root and soil pests. Thus far, all the chemical defenders, or at least all the writer has seen used, must be applied accurately and carefully, and none is completely harmless when applied directly to living root structures. For the most part, all of the compounds or groups of compounds mentioned here should be injected into the soil several days or a week or longer before the actual planting is done. The waiting periods are being reduced, and the goal of doing the most damage to harmful root pests with the least possible upset to adjacent soil lives is being attained with improving and oftentimes astonishing success.

Soil improvers, for which "root protectors" would be a better name, designed to inhibit the growth of specific root enemies, are already far more effective than even the more optimistic researchers would have guessed a mere ten years ago. They are also proving more and more valid in terms of business or profit. Methods of application are being similarly upgraded. Water solubles, such as copper sulfate, formaldehyde, allyl alcohol, and some of the mercury compounds, are being applied fairly effectively as drenches—also as liquid sprays for aboveground plant portions by way of which the root tissues can receive the treatment.

More and more of the solubles are being mixed in irrigation water or in pipelines and spraying machinery. Materials that are neither soluble nor volatile are being worked into the soil mechanically as powders or dusts. Tractor-powered and highly competent machinery is now available for applying them directly to root spheres. Manufacturers of farm machinery vie to bring out newer and better injection equipment, but there remains a most regrettable shortage of simplified, effective, and manually operated applicators for use by home gardeners and small farmers. In any case, the present vista of defending and improving root health by means of carefully compounded chemicals is brightening and broadening.

Not at all surprisingly, the most brilliant and promising findings in chemical means for strengthening roots and improving root health are taking cues from the almost unbelievable chemical feats performed by the roots themselves. This advance in doing what comes naturally to roots, even though it still seems nearly miraculous to mortals, is based on the living chemistry of roots.

The known gist here is that roots take from the soil of their own rhizospheres the moisture and minerals they need for their own growth and for the sustenance of their plants. The eventual products are our sustainers, the direct providers of hundreds and thousands of our vital needs. With their unrivaled chemistries, vegetative roots blend with magnificent care and still more fabulous precision the ingredients that they choose, and from them build sustenances for earthly life.

We now know beyond any real doubt that plant roots actually build together the amino acids that make the protein that sustains people and other animals (including insects) as well as plants. All these and multitudes of other necessities and life-improving luxuries, including spices, miscellaneous flavorings as well as all our principal food crops, resins, latexes and numerous drugs are on the book-long lists of life-sustaining goods that roots serve as prime suppliers.

In addition to making these products and thousands on thousands more, roots synthesize the amazing materials that regulate and direct the growth of their plants.

At least since 1937, plant scientists have been expertly certain that plant growth is regulated by hormones. We now know positively that roots manufacture many, almost certainly most, of the chemical regulators that direct and coordinate a plant's growth. For the most part, these growth controls are amino acids. The double blessing is that many can be effectively duplicated by people. The first to be extensively synthesized was indoleacetic acid. Other hormones retain enough of their secrets so that only an attempt can be made to synthesize them on the basis of similarly patterned molecular structures.

Three main classes of synthetic vegetative hormones are now being manufactured and used more and more beneficially for people. These are the gibberellins, the kinins, and the auxins. Each is known to be present and working in the root structures of a great many specific plants, and each has its own distinctive value.

Gibberellin, originally isolated and produced from the soil-dwelling fungus of that name, is now readily synthesized. When circulated through the roots and vascular structures of plants, it causes spectacular leaf and stem growth.

The kinins serve as circulation stimulants and capillary openers. One of their commercial uses is as a freshener, capable of keeping green vegetables crisp and unwilted for prolonged periods.

Auxins are used as regulants for fruit crops. Applied as leaf sprays—the prevailing method—they are used to prolong the fruiting season by retarding leaf shed or seed ripening, which usually causes fruit to drop. Or, early in the season when frost injury is threatened, they can be used to delay the opening of leaf buds and flowers.

There are still baffling areas in the astonishing realm of plant hormones that are principally produced by roots. For example, a given concentration of indoleacetic acid that strongly stimulates top growth is likely to retard root growth. The kinins, however, are becoming known and helpfully accepted as root inducers or stimulants.

The chemistry laboratory that, in all reality, every root structure is also has most remarkable placement facilities. As we have already noted as another item of common knowledge in the case of hundreds of species or genera of vegetation, roots can emerge from the stems of plants. For centuries, plant students were inclined to consider this an odd occurrence. Certainly, propagation of plants by stem cuttings is an ancient practice, dating back at least to Aristotle's time (around 384–322 B.C.).

Through the centuries the propagation of certain crops by cuttings has become routine practice. One simply cuts stem portions from the desired plant and places them in damp sand or other suitable medium. In due course rootlets appear. As any home gardener or orchardist knows, this routine has some impressive advantages over seeding. Rooted cuttings maintain the characteristics of the parent plant more accurately than do seeds, and various faults of seed propagation, such as embryo deformities and clinging fungal spores or insect eggs, are avoided. Growing time is shortened.

After P. W. Zimmerman's historic work of the late 1930s with plant hormones at the Boyce Thompson Institute for Plant Research in Yonkers, New York, plant scientists near and far began to discover that the phenomenon of stem rooting is directly related to auxins, including indoleacetic acid, which strongly stimulates the growth of roots. Synthetic versions of auxins, produced on a commercial basis, are now in widespread use. One is indolebutyric acid. When painted or sprayed on rose, grape, and other stem stocks, this and similar manufactured hormones cause roots to sprout promptly and vigorously on above-ground portions of plants. This, one of the truly great productive discoveries of our century, is already of epic benefit to plant breeders, fruit growers, nurserymen, florists, farmers, gardeners, and all who work with or earn their living from plants. The first credit is due to the enlightened teamwork of chemists and plant physiologists, who are learning more and more about living roots.

Applied at different strengths and at different times of the year, synthetic hormones can be used to regulate many crucial phases of a plant's growth. Comparatively simple acids such as naphthaleneacetic (its name is far more formidable than its actual chemistry), mixed into standard fruit sprays, are now keeping several of our most valuable fruit crops operable on a commercial basis by delaying or advancing ripening as desired.

For one example, spraying apple trees more or less regularly with a hormone mixture can extend the season of apple harvest by many days or several weeks, and the fruit retains good texture, color, and flavor until consumers can get the good from it. The commercial pineapple crop is now almost completely dependent on treatment with plant hormones that delay and extend the ripening periods so the harvest will not glut the markets and can be canned economically. Increasingly, as our knowledge grows, we are learning to use the hormone, the enzyme, and the living creative cell to grow sturdier, healthier plants which will give us finer products. The medium attainment stays the living root. The advance is being led by the root chemist, followed more and more closely and perceptively by the human chemist.

By studying roots with deference and imitating the competent work they accomplish, we are also gradually becoming aware of the living chemistry that sustains the different communities of plants—those that grow on sand dunes and the fringes of deserts, on rock-littered wastelands, in abandoned fields, at the margins of forests, and in established forests, meadows, and planted fields. We are beginning to see more and better ways of correlating the chemistry, biology, physiology, genetics, and other life sciences with what is often called climax growth. This is the ultimate plant community that continues to reproduce consistently and thereby to endure, more or less indefinitely. We are learning, too, that whatever the locale, the native plant community is the best of all laboratories for the study of roots and root chemistry.

More and more regularly with the passing years, plant scientists work in teams. During recent years the writer has been enlightened by interviewing seven of these study teams currently at work in North America. Among my admitted favorites are those based at the Boyce Thompson Institute for Plant Research in Yonkers, New York; those at the Shell Agricultural Research Station in Modesto, California; and others at Cornell University and the University of California. But the senior research team maintained by the U. S. Department of Agriculture in Beltsville, Maryland, is rather special. One keeps on encountering its veritable parade of first findings about roots. Heading the old reliables of the U.S.D.A. is J. W. Mitchell, a veteran chemist and plant physiologist, best known for his work in developing nonpoisonous weed killers.

The effectiveness of weed killers depends primarily on root functions. Most weed killers are synthetic vegetable hormones that upset the metabolism of specific plants (including many weed pests) and cause them to literally grow themselves to death. Working with Mitchell, at least as this is written, are P. C. Marth, a superbly competent biochemist; P. J. Linder, a deservedly renowned plant pathologist; and Margaret Montgellian, one of the most able laboratory technicians I have ever watched at work. The team as a whole and each member in his own right have made important contributions to our knowledge of roots.

Mitchell's historic and internationally used week killer, called 2,4-D (for dichlorophenoxyacetic acid), was produced exactly ten years after Zimmerman's initial work with plant hormones. Many other effective plant-metabolism influencers have been developed since then; quite probably at least a dozen more will emerge before this book does. But at least for the present these nonpoisonous herbicides are examples of what can be accomplished when scientists in various fields work together to learn from roots. As this is written, Mitchell and his team are seeking to learn more about the relation of growth regulators to root secretions, and about the chemistry of root-cell enzymes. The research is admittedly "open end," and, as often happens, something new has turned up that could profoundly change the prevailing concept of still another crucial aspect of plant physiology.

Students have known for a long time that the phloem or inner areas in both plant roots and stems serve as traffic channels for the delivery of nutrients, and that the fibers in the phloem also support the stem. Recent studies by Mitchell and his colleagues indicate that living cytoplasm (the protoplasm of the living cell) flows in a rotational course *within* these fibers, which until now were believed to have no other function than support. These research men found that the cytoplasm moves in the phloem fibers of bean plants, for example, about twenty-five times as far as it does in an ordinary cell. If it is proved that these fibers do serve as channels for plant foods, growth regulators, and other organic substances, we shall be another step ahead in man's understanding of the upward and downward flow of food materials in plants. So the gleanings of root chemistry gain flow and movement.

With the gains of light there are some momentous gains in the chemical relations and appraisals of heat. This, too, is a crucial phase of a crucial subject.

Almost certainly fire is man's oldest weapon against evil spirits within the earth. In 1947, in a hidden corner of lower Mexico, I had a rather lengthy glimpse into the Stone Age when I saw Indians performing their ancient ritual of purifying, or sanctifying, their garden-sized fields with bonfires. They were the White Robes, or Lacandon, who now live in the lush Lacanjas Valley of the great Indian state of Chiapas.

The Lacandon (around a hundred remain) are believed to be the last of the once-magnificent Mayas, or just possibly they are codescendants of the great Maya forebears. They still speak the Mayan language and worship in the ruins of ancient Mayan temples, chanting the same prayers and adhering to the old beliefs. They grow in their gardens what could well be the same varieties of corn, calabash, climbing beans, cotton, and tobacco that their forefathers planted, and use agricultural methods that have been handed down to them through the centuries. The fire ritual is part religion, part agricultural practice. Probably the treatment is not especially effective, except for the help the soil nutrients in the ashes give, because the heat does not penetrate deeply enough, but the age-old principle is essentially the same that motivates one of the newest of agricultural practices. I was deeply impressed by this fact, and by the absolute belief of the White Robes that the frontier of abundance lies beneath the surface of the earth, not above it.

In 1961 I followed a modern version of the fire ritual halfway around the world, in Australia. There, scientists of the Waite Agricultural Research Institute of the University of Adelaide—one of the world's truly great research centers—were demonstrating the use of steam in controlling the pathogens of root diseases. J. M. Warcup of the Institute is one of the most distinguished soil experts of our time and an outstanding authority on the heat treatment of soils.

Briefly, the Warcup thesis is that fungoid pathogens are the most serious menace to root health and that the best way to destroy them is to apply heat. If the heat incidentally kills the "good" fungi or decimates their ranks, the soil community as a whole will not be irreparably harmed. However, the parasitic fungi as a group are more vulnerable to heat than the usually beneficent, nonparasitic kinds. The temperatures applied must be completely controlled.

Along with appraising a thermally changed vista of soil chemistry, Warcup has accurately determined that when soil is heated to a temperature of 180 degrees Fahrenheit for as long as thirty minutes, practically all fungi die. Soil bacteria as a group can endure such a temperature and are likely to survive in soil temperatures below 160 degrees F. At a soil temperature of 120 degrees F., the fungi present can be listed as surviving even if discomfited; only the viruses are likely to be casualties.

A typical fungus test made by Warcup shows that in fertile soil thirty-one commonplace and harmful fungal species continue to survive a maintained soil temperature of 120 degrees F. From that point upward, the mortality rate rises sharply. Only eleven of the species survive at 130 degrees F. for as long as half an hour, and only seven of the thirty-one pathogenic species can survive a temperature of 140 degrees F. for half an hour. Seven species endure through 160 degrees F., but at 170 degrees F. only three survive. The 180-degree temperature, at which all fungoid species die, does not seriously damage bacteria, including the valuable saprophytes (which live on decaying matter), because they tend to colonize in a wider range of substrata.

The higher the moisture content of roots and soil, the greater is the susceptibility of living material to destruction by heat. It is generally agreed that for soil residents, as for other life forms, the specific cause of death from high temperature is the denaturing of proteins. When their molecular structure is altered, enzymes (which are proteins) can no longer function. If enzymatic action stops, all vital activities that characterize a living organism cease. Overheated soil organisms are thereby subject to immediate death.

There are three forms of heat that can be applied to root spheres: dry, hot air, hot water, and steam. The hot-air treatment is used mainly to destroy certain viruses or fungi by holding the temperature at about 130 degrees F. for several hours in a confined area or treatment bin. This is now a common procedure for safeguarding the roots of fruit-tree planting stock against virus infection prior to planting. With good equipment, even the most obdurate root viruses can be cooked out of the planting stock at temperatures that do not injure root tissues.

The hot-water treatment is most commonly used to treat seeds prior to planting. The seeds are kept in water that has been heated to a temperature no greater than 135 degrees F. for about thirty minutes, or in lukewarm water (around 110 degrees F.) for several hours. Cuttings and planting stock generally are treated briefly with comparatively hot water, or for a longer time with moderately hot water, to destroy harmful viruses.

Soil treatment with steam is now the most promising thermal procedure. The method is believed to have been first used in Germany in 1888, where it was known as the Frank method. During the 1890s it was popularized in the United States as the Rudd method. Early applications of steam to soil were too violent to bring good results; usually the soil was virtually sterilized. With modifications, however, the method endured as a more or less standard greenhouse practice.

Today, with the development of mechanized equipment, soil steaming has been extended to gardens, fields, orchards, and other commercial plantings. By means of it, various diseases of sugar cane, cereal grains, vegetables, ornamental trees, and fruit trees have been or are being brought under fairly dependable control.

The steam boxes now in fairly common use were invented in Adelaide, Australia, where the new heat techniques were perfected. They are portable boxes, fitted with pipe grids, which can be used to treat bulk soils in garden, field, or orchard. They deliver steam at a precisely limited temperature.

In the fruit-growing valleys of California steam injectors called Venturis are performing purification rites in fields and orchards. These are vats equipped to produce and apply to tilled soil aerated steam representing 2000 pounds of water per hour, at a uniform pressure of about 120 pounds per square inch and a soil-stabilized temperature of about 120 degrees F. This temperature is not harmful to plant roots and, if kept under accurate control, does not harm great numbers of soil lives other than the fungoid pathogens against which it is directed.

For field and orchard use, the aerated steam is introduced into the soil to a depth of eighteen or more inches by means of an implement called a steam blade. This consists of a parallel set of steam pipings twelve feet long which are fed by a butane-heated vat directly behind them. The equipment is drawn by a heavy caterpillar tractor. Lowered by means of a power take-off, the injector blade spreads steam through the soil. It can be raised or lowered as required. About sixteen inches of soil can be treated per minute. By means of controls, the temperature of the steam can be adjusted to bring about soil temperatures ranging from 100 to 212 degrees F. The highest temperatures are used only in limited areas where complete soil sterilization is desirable.

"Kill" temperatures, however, are being generally avoided. The current goal is the carefully studied and well-controlled application of temperatures that can do no more than temporarily revise the prevailing balance or imbalance of the intensely competitive flora and fauna of the soil—and even this with specific and usually temporary objectives. Authorities on the thermal defense of roots from soil pathogens stress the need for caution, accuracy of application, and more precise knowledge of the ways and needs of roots, rhizospheres, and biospheres. There is a consensus of opinion that the use of kill temperatures invites renewed contamination and is inadvisable for other reasons as well.

Meanwhile a more casual type of heating device is gaining popularity in farming circles. It is not intended to slay microscopic pathogens by the millions, and no complicated chemistry is involved in its use, but it does help the roots of plants in a number of ways.

The device is a heat-retaining cover, usually a very thin sheet of more or less transparent plastic such as polyethylene, which is spread on the hills or rows where seeds have been planted. The film keeps the surface soil a few degrees warmer than it would otherwise be, and roots and stems flourish lustily. In addition to warming the earth slightly, the cover protects the young plants from slashing rains, sleet, and hail. It also discourages competition by weeds.

The strip can be cut to any size or shape, and special tractor attachments are available that will lay the material, cut it, and even make holes in it for the convenience of emerging seedlings.

Aside from, or because of, various and sundry chemical reactions, the use of such soil warmers pays off in increasing harvests. In a recent experiment conducted by Michigan State University, careful records were kept. These showed increased yields of marketable vegetables as follows: eggplants, 31 per cent; muskmelons, 189 per cent (an average of 198 bushels an acre); tomatoes, 1180 bushels per acre instead of 507; green peppers, 1285 bushels as compared to 905; and so on.

Back of, usually beneath, the endless files of significant instances or timely examples is the clarifying certainty that the living chemistry of roots is subtly but profoundly correlated with the yet uncountable legions of soil lives that relate functionally or otherwise vitally to roots and their wondrous facilities as chemistry laboratories. Now in our still-toddling moves to intelligently explore this vast realm of the little-known we find cause for even more cosmic reflections: among others, that the prevailing, stiflingly expensive, politician-befouled "Space Age" is also very heavily dependent on knowledge of earthly soil life which we are only now beginning to acquire.

Until we have learned vastly more about the living chemistry and its interrelations within this first stronghold of life on planet Earth, how can we hope to comprehend life on other planets— which may very well be principally or wholly subsurface or subterranean? Querying another way, would it not be cogent, before spending the next X billion dollars of taxpayers' money to "probe" the moon or Mars, to at least learn the ABCs and two times-twos of this all-sustaining earth film beneath our feet? Thus far in the latter underworld, we are struggling somewhere between the B and C and sweating to attach a tail to the second two. This query is briefly inserted as still another suggestion-box contribution to NASA.

However marvel-crammed or however brow-crinkling the labors to learn more about the laboratories of roots, the gleanings and glimmerings thus far attained abound in hopeful shapes and foreshadowings. The hope-inspiring earth reading continues to stress the abilities of roots to fend for themselves with the magic of soil chemistry and the aid and comfort of the preponderantly good and largely invisible citizens of the living soil communities of which they are part.

As in most exploratory enterprises, not all the figurative cries of "Land ho!" are factually justified and not all sighted lands are directly connected with mainlands. Quite clearly, the foregoing holds true for genetic attainments in augmenting the ability of roots to defend themselves against the more conspicuous and better-known natural enemies, including thousands of "bad guys" among the millions of insects, or closely related animal lives; the hundreds of harmful fungi among a hundred thousand or so known species; the several scores of bad nematodes among the thousand of species now known, and so on.

Recently a Canadian plant breeder, D. H. Heinrichs de-

veloped several strains of the valuable northern forage, wheatgrass (*Agropyron*) that are highly resistant to grasshoppers,
heretofore their most destructive enemy. Dr. Heinrichs began
his work by crossbreeding grass varieties known to be exceptionally vulnerable to grasshopper damages. The ensuing hybridizing
enhanced root vigor and seemed to produce or activate genes
that enable the roots to contrive chemical compounds that
destroy the flavor appeal—to grasshoppers, but not to cattle,
sheep, or other grazing animals.

This man-contrived development bears on the great and multiplying capacity of root chemistry to aid in the fast-gaining
protective science of biological control. This working strategy
is based in important part on the entomologist's axiom that every
harmful root or plant pest has its own largely soilborne enemies,
insect, virus, fungal, or bacterial, which can and do infect, frustrate, or destroy it. The prevailing advances in biological control are marked by an unprecedented era of international cooperation in the defense of roots; the departments of agriculture
of the United States, Israel, India, Pakistan, Poland, and Spain
have already brought together one of the more able research
forces.

Closely correlated with biological control methods are intensified studies of genetic and chemical traits that serve to defend
certain plants, particularly their root systems, against insect attack. A good example is sweet clover. In the middle 1960s a
weevil was attacking the roots of sweet clovers to such an extent that, valuable as these great legume forages are as crops,
they were rapidly being abandoned. The apparent remedy lies
in plant breeding—specifically, the crossing of field varieties with
a common wild clover that for some reason the enemy weevil
(*Setona cylindricollis*) avoids. Here again, scholars believe that
some form of root chemistry gives the wild clover self-developed
protection and that, by crosses, the protection can be extended
to field varieties.

Viruses, mainly soil-based species, are another powerful

weapon in the war against destructive insects. Dozens of them have been identified as enemies of one or more pests. For example, a virus of the polyhedron type, *Heliothis zea,* is known to infect and destroy the cotton bollworm (which is also the corn-ear worm and the tomato fruitworm), and a similar virus, *Heliothis virescens,* is known to destroy the tobacco budworm. Virologists have proved that these and perhaps hundreds of other species can be mass produced successfully in laboratories.

The U. S. Department of Agriculture has recently established its first official quarters for the mass production of such viruses. The laboratory, in Brownsville, Texas, is under the direction of an exceptionally imaginative entomologist, C. M. Ignoffo. His most ambitious goal, shared by his colleagues, is to develop nonpoisonous insect killers and place them within the reach of farmers, gardeners, and plant enthusiasts everywhere. Mass production begins with feeding the virus polyhedra to the larvae of the plant-destroying insects. Once within the larvae the polyhedra multiplies fabulously—up to 10,000 times its original volume. When the larvae die, they are processed to obtain a virus suspension that can be sprayed on plants. In the case of the antibollworm suspension, the virus content of 100 infected larvae is sufficient to treat an entire acre of plantings. Moreover, the virus material can be preserved by drying or freezing.

When the pests feed on the treated crops, they become infected with the virus and transmit it to others of their kind. A notable feature of insect control by the use of virus materials is that the viruses usually infect only their specific targets. They seem to cause little or no damage to beneficial insects or to animals.

Back of man's ingenious advance in helping roots and root chemistry stand against the natural forces that would destroy them are the unstinted and largely invisible and unidentified good citizens of the living soil community around them.

A reassuring instance recently came to light when two agricultural students, one at the Imperial College of Tropical Agriculture in Trinidad, the other at the University of Maryland, reported independently that shredded straw mixed into the soil seems to repress the fungi that cause root rot in bean plants. At about the same time, and again independently (each was working on his own time and in his own garden), two microbiologists of the Agricultural Research Service of the U. S. Department of Agriculture were pondering the apparent fact that adding humus material to the soil frustrates numerous harmful fungi. In their first experiments the U.S.D.A. men identified eighteen species of soil fungi as pathogens of common root rots. Acting on simultaneous hunches, the two microbiologists, C. B. Davey and G. C. Papavizas, decided to loosen the soils by working partly rotted straw into them. A few weeks later they planted various garden seeds, including beans, peas, and cucumbers. The seedlings grew lustily but the soil fungi seemed oddly inactive. Only four of the eighteen previously identified fungal pathogens could even be traced. On painstaking investigation, Davey and Papavizas found the reason: it was the presence of a genus of purplish soil bacteria called *Pseudomonas*. These bacteria, living close to the plant roots, produce an antibiotic that proved fatal to fourteen of the eighteen fungus pathogens present and, one gathers, quite sickening to the rest.

Adding the shredded, partly rotted straw had helped the purple bacteria by raising the carbon-to-nitrogen ratio to a level especially favorable to them. To state it in numbers typical of bacteria counts: in acid soil where the ratio of carbon to nitrogen is fifty to one, the number of bacteria that are effective against the root-rot fungi averages about thirty million for each gram of soil; in alkaline soil where the ratio is the same, the bacteria count is substantially lower but there is still effective protection. Heavy applications of commercial fertilizer with high nitrogen factors reduce the population of the beneficial *Pseudomonas* bacteria and encourage the disease pathogens.

This discovery fortifies the views of dedicated organic gardeners. It also bolsters the theory that the best way to improve the health of plants is to aid them in keeping up their own natural defenses. There is sanctuary in the soil, where roots prevail and rule; there are ways in which people can serve as benefactors as well as beneficiaries of roots, and the amazing mystique of root life grows ever more interesting as one begins to heed and study it.

11

ROOTS: BATTLEGROUND OR SANCTUARY

Many sciences, professions, and industries generate their own backgrounds of skepticism. A competent detective, for example, is likely to have a line-of-duty skepticism that augments the doctrine that all knowledge is uncertain with the acceptance that the motivations for crime are omnipresent. To a marked degree, insurance is an industry of skepticism. With the indispensable help of investigators who, however discreet, are professionally addicted doubters, insurance stays mighty by "protecting" people against the possible but the mathematically improbable.

Usually, the competent pathologist is a man or woman of scientific disciplines merged with the intellectual traits of the enlightened skeptic. On this substantially reasonable basis the adept soil pathologist is strongly disposed to see root spheres as first-line and incessant battlegrounds where living roots are beleaguered unendingly by frequently invisible enemies in great part more nearly immortal than roots themselves.

As a rule the currently known or strongly suspected root pathogens do not just go away. Although the root afflicters are admittedly an extreme minority of soil lives, they are not generally disposed to yield their space or otherwise give way to less antisocial lives. This reinforces the pathologists' awareness that the life film called soil is in fact a usually multiplying, typically unstable, ever-teeming mob of diverse tenants competing for living space and, figuratively speaking, for the food platters. The competition for space and food is perennial, perpetual, and, from all appearances, interminable.

Dr. William Snyder of the University of California (Berkeley), who heads the nation's largest and one of the most respected faculties of soil scientists, makes the very rational point that as more acres of soils are brought into more intensive cultivation, and as soil films continue to grow thinner from the ravages of erosion and tillers' abuse, we must expect the wrongdoers in the root spheres to grow more numerous and more aggressively damaging.

The position is powerfully supported by persistent and forehead-crinkling evidence that both the varietal range and the combined power of root pathogens are steadily if stealthily increasing. This seems to be the unwelcome truth, quite regardless of epochal advances, both recent and current, in repelling the onslaughts of particularly obnoxious root diseases and specific fungal, insect, bacterial, and virus enemies of root structures. It is true despite the improving musterings of chemical benefactors of root health and the already magnificent odyssey of building up root strength by means of vegetative genetics.

With willingness to salute and welcome the new argosies of attainments and discoveries, many pathologists believe that, as of today, prevailing levels of root health, particularly those of numerous cultivated crop plants and forest trees, still continue to diminish.

For the most part the more compelling reasons for this do not relate preponderantly to the so-called higher orders of fungi or insect pathogens. As already suggested, the great dark persistence of root diseases lies in the lingering twilight world of what are conveniently termed the lower orders of root pathogens. At least in formidable part, these are the harmful, disease-inciting single-cell lives, bacteria, viruses, and the lesser-known and more harmful fungi. These remain principally in the fold of the more mysterious soil lives, mysterious in terms of their tininess—in many instances, thousands on thousands could camp on a pinhead; in histology or tissue structure; in mutation or direction of progress or, alas, mischief. Like so many other life forms, these microscopic or submicroscopic undesirables apparently have never heard of the Birth Control League, have never read Senator Fulbright's speeches on that subject, and, one can infer, have never had so much as a virtuous urge to practice celibacy or to subscribe to Malthusianism or, indeed, to differentiate between the Reverend Mr. Thomas Robert Malthus and the Reverend Dr. Norman Vincent Peale. (Actually, of course, many of them reproduce asexually by means of fission.)

These undersized, still preponderantly mysterious plaguers or threateners of root health throughout the tenanted earth, from tundra edges to the lushest equatorial valleys, are not abstaining from self-perpetuation and, unlike the sung-about old soldiers, do not just fade away. Instead, the host of the tiny and too-little-known continue to compete and conspire to make root spheres ever more recognizably to the battlegrounds for a better world of roots and root dependents, including man.

Thus, the incentives for learning more about roots and, where possible, helping them help themselves are at an all-time high. Never before has knowledge of roots been of such momentous consequence to people. As we are so graphically reminded by the United Nations Food and Agriculture Organization, some

two billion people, more than 60 per cent of all now living and
majority citizenries of more than a hundred nations and three
continents (Africa, Asia, and Latin America), are currently
going hungry or living with food deficits. Of these hungry or
hunger-threatened two billion, at least 600 million are farmers
and therefore especially dependent upon the hardiness of roots.
Necessity has long been regarded as the mother of invention,
but the role of this mother has never before been of such crucial
importance to so many billions of people. National, international,
and intercontinental food needs not only require but demand
the immediate multiplication of all kinds of edible harvests. In
turn, even the minimal fulfillment of food needs postulates the
continued improvement of root health. This cites the cause and
the places of the primary, globe-circling battle or, if one pre-
fers, crusade against hunger.

The first elements of battle plans include information about
the enemy and about the supporting, friendly forces. Roots are
the primary supporting forces. They cannot be adequately mar-
shaled until they are much better understood. The great life
sciences are striding ahead in attaining this understanding,
which is prerequisite even for locating the battleground. For
the science-trained realist, upgrading root health is the first
maneuver in this all-crucial battle. It requires ably correlated
knowledge and appreciation not only of the imperiling enemies
but of the implicit competence and strength of roots and root
spheres both as nature's foremost battleground and as its most
encompassing sanctuary.

The knowledge tends to place the knowers and learners in the
role of sorcerers' apprentices who participate in an every-amaz-
ing processional of change. For whatever their other characteristics, the multitudes of lives that make up the soil are never
static; they are forever going and coming, always influenced by
ever-changing factors such as moisture, heat, tilth, and aeration.

Certainly soil moisture is among the more powerful and persistent forces of change. As roots grow and penetrate the soil,

they permit the leaves and other above-ground portions of their plants to take up more and more of the soil water supply and otherwise reshape its climate. Among the related factors are size and density of plant populations, rate of leaf respiration, extent and density of leaf shades, and, of even greater decisiveness, the prevailing rates and patterns of root growth. As earlier intimated, the last-named is also the primary decider in soil aeration; when aeration is increased, many kinds of bacteria, fungi, and other soil lives suffer (or enjoy) population explosions, while others, including the anaerobic bacteria—those that live without air—usually diminish. Leaf growth, meanwhile, reduces the amount of sunlight that reaches the soil surface, thereby causing a decline in temperature in root spheres and otherwise changing their climate patterns.

All the foregoing belongs in the for-better-or-worse department, but effective exploitation for the benefit of roots is clearly among the great agronomic challenges present and future. Among the already established vantage points for hardier roots (and better-fed people) are the exudations from roots.

This is an area of very special challenge for soil chemists, biologists, plant breeders, and, quite directly, practical farmers or gardeners. The now absolute knowledge that every plant species has its own distinctive root chemistry stresses the advantages of having a greater variety of valuable crop plants for meeting or correlating the chemical cycles of a given soil.

We still need a longer, more versatile list of available and adaptable crops. We have already noticed some of the highlights and specifics of developing new crops or better varieties or strains of already established crops. In passing it seems fitting to note that along with the many new crops that are now needed are a number of once important food crops which have now been largely abandoned. Examples include such excellent food sources as salsify, barley, buckwheat, artichokes, pomegranates, and many others. As people and the need for feeding people grow stronger, the need for a great many new or renewed crops keeps gaining urgency. Expert plant explorers, including very able search teams employed by the U. S. Department of Agriculture, estimate the attainable number of available new crops at high in the hundreds. A recent survey directed by Dr. Robert Harris, of the M. I. T. Biochemistry and Nutrition Laboratories, recently located and made expert nutrient analyses of 246 valuable food crops in Central America alone, in a land area barely one fourteenth the size of the United States.

Another prime benefit of developing and adapting new crops is the opportunity, indeed the obligation, to advance the study and knowledge of roots. This heartening truth became impressively clear many years ago; during the 1930s we began to notice that rootwise the failures in this pursuit can be even more educational than the successes.

Back in 1936, F. A. Haasis, one of Holland's most brilliant soil pathologists, set out to save the iris-bulb industry of his homeland from ruin by fungoid diseases. By doubling as an extremely cogent chemist, Dr. Haasis identified the bulb infection as the blighting fungus *Sclerotium roffsil*. Haasis developed an effective control, but no sooner was it in use than a formidable and until then all but unknown horde of soil pathogens swarmed into the treated soil. Unwittingly, he had contributed to the upset of natural balance of the soil lives, thereby admitting root-sphere villains that turned out to be far more damaging than the first "most wanted."

At about the same time, Britain's renowned soil chemist I. S. D. Gibson, who had earlier developed an organo-mercurial seed treatment for controlling the crown rot of peanuts, worked out an ethyl-mercury-phosphate compound for pretreating pine seedlings against the pathogen that causes the so-called damping-off disease. Here again, the fungicide served its specific purpose, but the kill compound seemed to have opened the way for a host of other pine-root diseases.

Gibson has continued to point out that a biocide or other toxic agent applied to soil without a precise foreknowledge of its reactions may indiscriminately kill both the harmful and helpful soil organisms; that the introduction of chemical killers involves some degree of peril, but the best root defense is to adapt plant species that best fit the soil in question.

The same period marked the emergence of the Weindling hypothesis regarding the inherent pathogenic relationships between different races of soil organisms. Emmett Weindling demonstrated quite impressively that the properly dreaded race of fungi, the *Rhizoctonia*—properly dreaded because it includes several of the most ruinous of the root rots—has its own potential checkmates of natural enemies in the soil. One of the most effective and widespread is the genus or family of fungi called *Trichoderma*. Very acid soils that are severely infected with the root rot can be treated and frequently "cured" by introducing the *Trichoderma viride*. Incidentally, the reason why the panacea usually works superbly in acid soils but not in nonacid is one of the multitude of mysteries which relate to living soil chemistry. But Weindling's discovery served as a particularly revealing flash of light that shows the way for building up one of the more promising defenses against famines or mass hunger.

A related revelation grows out of pioneer findings of one of the more renowned U. S. Department of Agriculture pathology teams, T. W. Millard and J. S. Taylor, who back in 1927 set to investigating the ruinous root scabies caused by another big and notorious fungus family, the *Streptomyces*. In a comparatively short time Millard and Taylor demonstrated that the very harmful root scabies can frequently be controlled merely by plowing under a cover crop of rye. There is powerful evidence that an abundance of organic material can enable soil lives to fend for themselves against what are otherwise ruinous enemies.

The hope-nurturing, people-saving point here is that, regardless of the ruthlessness of root diseases and their abilities to upset natural balances, the living soil maintains the precious nucleus of a biological basis for reducing or curbing the root injuries inflicted by soil pathogens. That, in turn, can often be advanced simply by choosing and cultivating another species or genus of crops. The going term for this potential of equilibrium of soil lives is "biophase." A decline in equilibrium or biophase competence is often noted in soils that are planted year after year with the same crop. Thus, the readily attainable feat—or at least it ought to be readily attainable—of changing the crop can frequently affect the soil community most favorably. But there is no one sure-bet, right-the-wrong crop or plant species for any or every imbalance of soil lives. A fairly sure cure for a particular type of soil may succeed promptly in improving the supply of nutrients available to roots and the populations of favorable soil organisms. But the same "balancer" in another soil structure may fail abjectly. Evidently a primary need is to learn much more about this biophase and to hold in readiness a much greater variety of food-provident or other economic plants as possible benefactors of existent or oncoming imbalances of soil or root-sphere population. Never before has a greater, better-studied diversity of crops been so crucially important to land, food production, people, and, most basically, to better root health.

Here again, the absolute indispensable is man's willingness to learn. At long last, after hundreds and hundreds of generations, man now is beginning to learn at least primerwise what soil is, what roots are, and which soil zones are most important to root life.

The more crucial of these, as one keeps noting, is and has to be the rhizosphere, the region closely surrounding the roots. The next in life-sustaining consequence is the biosphere, the more general life zone surrounding the rhizosphere.

Living chemistry grasps hands with physiology as the living presence of roots in the rhizosphere encourages the multiplication of the formidably named mycorrhizas—root-clinging fungi that are know to be indispensable to the life and health of many plants. Also in the rhizospheres are the prime makers of soil, the multitudes of microorganisms that decompose available proteins, fats, and carbohydrates for nourishing roots and the plants they support. In this same vast and wonderful echelon are the hardly less essential converters of the life-sustaining minerals that roots must collect—phosphorus, nitrogen, iron, sulfur, molybdenum, and many others.

It would be a great comfort to all soil students if some kind of zoning ordinance could be passed so that the bacterial mineral converters, including such all-necessary races as *Rhizobium*, *Nitrosomonas*, and *Azobacter* could replace or quietly stamp out the undersized imps such as the bacilli that conspire to lock away the mineral nutrients that all roots apparently require. But even as with people, intermingling of good and bad seems to be the preordained nature of soil communities. Such is the meekly accepted bitter tea in what we now perceive as the sanctuary of roots.

Bitter tea or no, and failures be damned, the deft and doing scholars of soil continue to learn and prove that at least some of the habituated violators of the sanctuary can be routed or effectively reduced by means less dangerous than chemical counterattack. The "shotgun method" of blasting soils with lethal fumigants is being replaced by more effective soil improvers. Though they are still experimental, the list of such improvers continues to grow.

We have already noted some of these briefly: synthetic growth hormones, metabolism stabilizers and upsetters, growth stimulants and retardants, specialized biocides, and other preparations designed to induce a desired response in roots, rhizospheres, and vascular systems. A research coordinator with the U. S. Public Health Service tells me that according to his records only about two would-be soil improvers in every thousand tried ever reach the stage of commercial manufacture. Even so, about 300 new compounds hopefully labeled as soil improvers appear on the American markets during an average year—60 to 70 per cent of them invented or developed in the United States.

The agricultural research laboratories of Monsanto in St. Louis provide an impressive view of this comparatively new billion-dollar industry. Here, any day in the year, about 2000 soil and/or root improvers are being devised, compounded, and tested. By averages, about twenty-five of these will survive the first year of practical field tests, and of these, four to six will survive the final year of seed tests, toxicity tests, manufacturing tests and trials on state and federal experimental farms. The next step for the survivors is official licensing by federal and, as required, state agricultural agencies.

Licensing is usually preliminary to two or three years of additional testing on selected private farms. After this comes commercial manufacture of the product and selling—a venture that often fails. Worthwhile products sometimes fail because of the high cost of their ingredients. At least in the United States an unwritten law seems to stipulate that no soil improver has a dependable commercial future unless its use repays the grower by about seven to one. It is also more or less required that a successful improver contribute to harvests of better flavor or more attractive appearance than could otherwise have been grown. It would seem that all these requirements would discourage the soil-improving ventures, but it does not work out that way. Since the late 1950s the development of root-benefiting compounds has definitely been big business in the United States and Canada, the United Kingdom, most of Western Europe, the Soviet Union, Australia, New Zealand, and more recently, through extreme duress of food needs, India.

In a broader sense, the soil improvers are extending further and further beyond the rosters of beneficial chemical compounds. They also include the great fundamentals of improved water absorption and enrichment by fertility—factors that are provided or underwritten by the age-old working partnership of earth, rain, sun, and air.

It is now common acceptance that roots in rich soil (soil that contains active organic material of around 2 per cent of volume) requires about one fourth the amount of rainfall or irrigation necessary for poor soil (with, say, 0.5 per cent of active organic material). One of the best tried-and-proved means for raising the "living fertility" and attaining the best use of the water supply within the root zone is the planting and subsequent plowing under of crops such as rye, vetch, cowpeas, clovers, etc. Such "green manuring" serves to strengthen roots by means of improved soil structure and better-sustained soil lives. The procedure aids the increase of beneficent microorganisms, and there is better than an off-chance that the improved nutrition may divert some evildoers from trespassing on or into plant roots.

The sagacious combination of green manure and chemical fertilizers is gaining favor as a means of increasing root vitality. One particularly effective procedure is the autumn sowing of rye grass and vetch, or some other appropriate grass-and-legume combination, and the application at that time of a light dressing of chemical fertilizer. The cover crop is turned under the following spring with a disk harrow or shallow plow, and the rest of the yearly quota of commercial fertilizer is applied. This routine is designed to benefit roots by raising the organic factors in the soil to a becoming high while maintaining a favorable soil structure and conserving the water supply.

Meanwhile there is a trend toward making a better appraisal of plants in relation to soil temperatures. Better root health and improved production are the continuing goals. Plant breeders and physiologists are learning to identify and chart temperature ranges that are suited to a given crop yet distinctly unfavorable to its root enemies. To note just one entry, it has been found that many of the root pathogens that afflict tobacco do not harm hot-country varieties that are bred to thrive in soils with, say, a mean temperature of eighty degrees F., instead of around seventy-five degrees.

Soil warmth or lack of warmth is inevitably related to moisture. This opens the way for still another strategy for defending roots. Since all soil lives require water, it follows that the water supply can be manipulated to destroy pathogens. Roots can

endure thirst far better than most of their pests can; by controlling irrigation accurately, some harmful fungi and insects can
be "thirsted out." Similarly, afflicted wetlands can be drained.
As a rule, the disease causers cannot adapt to abrupt changes
in the water supply.

Certain minerals also tend to influence soil temperatures and
discourage attack by fungi and insects. As a very general rule,
the count of soil pathogens goes down when the lime content
goes up. Adding basic food chemicals, particularly nitrogen
and phosphate, and organic materials that improve the tilth of
the soil while adding nutrients is frequently helpful, too.

Healthy roots improve soil aeration, which in turn is one of
their better defenses. Sun covers, too, are frequently beneficial.
Contrary to the texts of some of our widely used schoolbooks,
not all fungi breed best in the dark. Some of the most ruinous
of fungoid diseases, including many of the rusts and smuts that
damage grains, flourish best in bright sunlight. Even partial
shade serves to discourage them. The use of trees and shrubs
to provide partial shade for smaller vegetation is a grand old
tropical strategy that deserves to be studied in terms of temperate-
zone plants. If coffee bushes were not shaded, we would probably not be drinking coffee today; the voracious fungi that attack
their roots would have long since seen to that. And many other
useful plants seem to need more shade than their own leaves
provide.

An exciting new development in soil research goes far beyond
prescriptions designed to improve soils that are already producing crops. Its goal is to employ the roots of succeeding species of
vegetation to cleanse, build, purify, and eventually make suitable for intensive cultivation many kinds of marginal and non-
arable soils. The premise of the soil chemists is that through
the ages roots have done much to build soil and make it increasingly productive. The deduction is that by understanding
this and studying the intricate angles of the problem, man can

help roots do an outstanding soil-building job in desolate areas.

Some call the developing technique the Feichtmeir method in honor of Edward F. Feichtmeir, one of California's distinguished practical pioneers in using roots to build soil. The object of the method is to build up the so-called badlands, including waste-lands where mineral accumulations such as salt and gypsum dis-courage vegetation. The first step is to plant hardy types of primitive vegetation such as low-grade grasses—the so-called scratch grasses or three-awns. These are determined clingers with valiant if undersized root structures that can endure drought and unfriendly soils. After the primitives have rooted, the land would be seeded with indigenous perennial grasses or excep-tionally hardy imports such as wheatgrass, Dallis grass, Middle East bluestems, and, in time, buffalo grass or Bermuda grass.

Once the perennial grasses are established, even if the cover is very light, the battle is won. After their roots implant or-ganic matter and open the soil to air and water, the next step is to sow by "scratch planting" (using light harrows or cutting rakes) any adaptable legume crop that will grow there. After ten or twenty years of preliminary soil building by the grass roots, it is usually possible to introduce a hardy legume crop to build up the supply of organic nitrogen. The kudzu vine and mountain lilac (*Ceanothus*) are good bets, but the great forage legumes—selected clovers and alfalfa—are the best. They implant organic nitrogen at root level, and in time they make a cover suitable for grazing. Eventually, at least in some instances, crop-land of fair fertility is achieved. A conversion operation of this sort usually takes from twenty to a hundred years and is there-fore best attained by government agencies or long-lived cor-porations. However, some of the most promising work done on saline and gypsum wastes in California has been done by pri-vate enterprise.

The Feichtmeir technique and those similar to it are also re-vealing new and significant facts about root life. For example,

Feichtmeir and his colleagues are producing evidence that plants able to develop a high carbohydrate level in their root bark or cork cambium are more than averagely resistant to the fungi that cause root rots. Apparently soil microflora that are resistant to the fungoid pathogens are attracted by the higher carbohydrate level. For reasons not completely understood, these levels seem to be highest in dry years and in low-rainfall areas.

Concern for the welfare of roots has taken researchers deep into subsoil strata. Compacted areas, called hardpan, seem to be increasing relentlessly on tilled land. Sometimes hardpan can be broken by a once-in-a-lifetime deep-plowing operation, but merely breaking it up is not the ultimate solution. Often the real cause of the trouble is the presence in the soil of minerals that are injurious to roots, especially their growing tips.

Superb research by the Auburn (Alabama) Agricultural Experiment Station has recently shown that deeply implanted aluminum is a prime offender. In nonacid soils (where the metal remains insoluble) little harm is done, but in acid soils, after prolonged cultivation, the aluminum becomes irritating or even deadly to the root systems of certain plant families. Cotton is one example. Other crops, including soybeans, do not seem to be vulnerable.

Plants that are vulnerable to aluminum toxicity are affected in their root cells. The nuclei continue to divide, but the new cells fail to develop walls. Acid-soil aluminum thus emerges as the most ruinous of cotton enemies. Manganese in acid soils is also damaging to many types of vegetation, and there may be still other lower-soil minerals that in combination with soil-engendered acids seriously injure roots.

Liming, a time-tested method of correcting soil acidity, is being used increasingly today in efforts to help roots help themselves. Lime is easily applied to the topsoil, but how to get it into the subsoil where the hardpan is remains a problem. A better technique than extra-deep plowing is needed, and chemists, physicists, engineers, and others are urgently invited to make suggestions.

The life-sustaining saga of roots and the ever more crucially helpful sanctuary of roots are inevitably and most vitally inter-related with plowing. As a long-time plowman for better or worse—as I now reflect, mostly for worse—this writer frequently ponders the theme and extent of the plowman's folly: the in-cessantly disturbing, violently upsetting, never-ending, befud-dled, ghoulishly well-intentioned perpetuators of mayhem of roots.

When my almost daily mail arrives, bulging with what is charitably known as junk mail, I sometimes mull over the prop-aganda tracts or money solicitations pertaining to providing plows for the quaint but hungering Petootians or extolling the noble humanitarian philanthropy of such-and-so foundation which has recently granted $319,011.08 for equipping and training the simple Quagohapoo Indians of Chichicastenango to plow the living daylights out of the eroding remnants of soil on their mountainside milpas—or garden fields. But I am not really convinced that the real quest is not for, say, $419,011.08's worth of publicity for the such-and-so foundation or some tax-exempt fund for whomever is proselytizing for plows. For a great deal of the arable earth does not need plowing, and tens of millions of effective farmers have never owned or used or wished to use a plow. Any thoughtful study of root structures and root needs shows the disadvantages as well as the advan-tages of plowing and suggests why the former sometimes out-weigh the latter.

When John Deere, a hulking Vermont blacksmith, invented the moldboard plow in the late 1830s, he thought he had really "done it." No longer would the prairie—with its deep, tough sod—frustrate the would-be farmer, and in the valleys the rich earth would be turned with ease. The story goes that Deere had been hammering out spades and hoes as a thirty-dollar-a-year ap-prentice and had managed to save up seventy-seven dollars in cash. With this backlog he headed west with, as he wrote, "a wild antic notion" in his head. The wild antic notion was "the making of a self-balancing smoothed steel plow which . . . can break through clay or mud prairie without fouling like other turnplows."

According to folk say, the roving Vermonter picked up a circular steel saw that had been abandoned by a roadside sawmill, borrowed tools, and went to work. He "beat the steel plate butter smooth," then shaped it into the vertical siding and cutting share for a turnplow. In smithy talk a "moldboard" means a curved metal plate attached to a plowshare. John Deere reputedly made the beam from a fence rail and shaped the handles from a prong-rooted sapling. The fiendishly effective sod turner was patented in 1839, only five years after the patenting and eight years after the first tryout of the McCormick grain reaper. But "Deere's Moldboard" was soon to outdo the reaper as a dominating farm implement.

A history-making orgy of plowing—and we have been inveterate plowers—followed in the United States and much of the Western World. The prairies were breached; hillsides, mountainsides, and valleys were turned; and vast areas were planted for the first time. Wind and water erosion stripped away the topsoil of rich farmlands and crops declined. Millions of acres of semiarid prairieland, bereft of protecting grasses, became deserts of windblown dust when droughts came, and during the rain cycle the rich soil of the valleys was flooded into the sea. Roots were sick when the land had been overplowed. We know today that even minimal cultivation increases soil acidity, encourages pathogens, and dissipates needed root foods, such as the amino-acid nitrogens. But our immediate forefathers heeded not. Perhaps never before in the history of man had the sanctuary of roots been assaulted with such vehemence and violence.

Almost exactly a century after John Deere had put together the plow that was to break the Great Plains and tens of millions of acres beyond them, a hard-working farmer and market gar-

dener from Oklahoma branded the moldboard plow as a prime
ruiner of land. With calm, bucolic fury Edward Hubert Faulkner
wrote a book lucidly entitled *Plowman's Folly*. Despite the rec-
ommendations of several distinguished scholars who had read
it, the manuscript was kicked about for several years. Finally
in 1943 the University of Oklahoma Press, under direction of
Savoie Lottinville, valiantly published the apparent dud. The
book generously rewarded its author's and publisher's persever-
ance, sold in all about 2 million copies and continues to prove
itself the most successful of some 1300 titles, by some 800
authors, thus far published by the University of Oklahoma
Press. The book created a furore in farm circles and impressively
influenced prevailing trends of agriculture. Faulkner's may well
be the most important farm book of the century; certainly it is
one of the most convincing.

As an experienced farmer and market gardener, principally
in Oklahoma, Faulkner stated the ever-demonstrable tenets that
deep plowing opens the soil to violent erosion, dissipates soil
water, ruins capillarity, buries organic matter beyond the more
effective root reach of many valuable crop plants—at least at
the time when they need it most—perpetuates more weeds than
it destroys, fouls up soil aeration, and, in general, ravages root
systems.

With considered emphasis, the deep-believing Oklahoman de-
clared that the moldboard plow should be outlawed and replaced by the disk harrow, or a similar cutting harrow or
"scratch" tool. Used intelligently, any one of these would return
cover crops, surface litter, weeds, and leaf mold to the rhizospheres of crop plants. Faulkner further recommended the use
of rollers to "staple down" the surface soil and give it maximum
protection against wind erosion. He pointed out that injury to
roots is unavoidable when one plows deep, and that if it must
be done, deep plowing should be regarded as a once-in-a-lifetime operation. The defense of roots and their spheres against

the ravages of conventional cultivation has rarely if ever been stated more ably. *Plowman's Folly* is still worth reading.

Today experts agree that overcultivation is the number-one bane of professional agriculture. Again consider corn, our major field crop. In my boyhood we planted corn by hand, hoed it at sprouting time, dutifully replanted, subsequently and dutifully "plowed out the middles" five times—no more, no less—during the growing season. The final two or three plowings must have been mayhem to the roots.

Recently, able investigators have proved beyond any faint shadow of doubt that the fourth and the fifth, frequently the third, and sometimes even the second plowings do more harm than good. Recently, experimenters at the University of Illinois, Texas A & M, the University of Maryland, Rutgers, and other testing farms have demonstrated that excessive cultivation can cut harvests by as much as 35 per cent. Yet weeds in cornfields must be controlled. Fortunately the midcentury development of chemical weed killers now makes this a safe and simple operation. At the New Jersey Agricultural Experiment Station, now allied with Rutgers University, experiments by Ming-Yu Li and W. F. Meggett show that a plot of 100-bushels-an-acre cornland will produce around 130 bushels an acre if weeded by hand. Weeds, of course, reduce the harvest. Experiments at the University of Illinois showed that even a three-inch strip of foxtail grass in the rows reduces the corn yields by as much as twenty-six bushels an acre. The Li and Meggett experiments prove that the yields drop as much as eight bushels an acre when weeds are permitted to grow for two weeks. So there is still a case for the Man with the Hoe and for the Nimble Boy on Hands and Knees Who Kept the Garden Free of Weeds. But there is momentous irony in the fact that during the century just passed, the plow has been the most formidable enemy of roots.

In a limited way, man has plowed the land for centuries, but there is similarly valid evidence that for hundreds of years before the simplest version of a plow was invented he produced great crops. And today—in Asia, Africa, much of Latin America, and the South Pacific—there are hundreds of millions of farmers, many of them highly proficient, who have never used a plow. Some have never even seen one.* Nor was the plow known to the pre-Columbian Indians whose numbers unquestionably included some of the most able farmers then living—or greatly used by the first European settlers of North America.

At present the pace-setting majority of designers and manufacturers of farm machinery are bolstering the thesis Faulkner propounded in *Plowman's Folly*. In great part they concur in his views and welcome the chance, as a British manufacturer puts it, "to act sane for a change." The tide has turned. "I'd frankly like to put a sign plate on every farm tractor," an executive of International Harvester recently told me, "with this query on it in boxcar letters: 'Is this trip really necessary?'"

Today farmers, gardeners, and orchardists in the United States and many other nations are pioneering tractor-powered machines that in a single operation till the soil, apply fertilizer and other treatment compounds, and plant the crop. As far as we now are concerned, any deep-cutting plow is immediately suspect. At long last, American and many other tillers and lovers of the earth are converted to the idea that soil really is for roots. Among them is one E. R. Poynier, a mechanical engineer and farm-machinery designer whose developments of cultivating equipment are now used in Canada, Australia, New Zealand, most of Europe, and the United Kingdom as well as in the United States. Rumor has it that some of the superlative new farming machines of Soviet Russia look and move remarkably like Poynier creations, as do some few in the Greater Peking Sphere of Prosperity. Poynier says: "Good farmers throughout the world are eagerly accepting the technology to

* The United Nations Food and Agriculture Organization estimates that approximately 100 developing countries in the present-day world have a total of around 600 million resident farm people of whom a probable 300 million do not use plows.

minimize cultivation. In every principal farming nation I know, the present cultivation goal is to try to do the crop the most good and its roots the least harm."

In many other ways the sanctuary of living roots is on the gain because of cogent acts and findings of people. Another redeeming area of advances relates to soil temperatures and the development and uses of thermal devices that help safeguard and improve root life. The natural-history background is essentially this:

We know that plant roots are extremely competent in measuring and utilizing the amount of heat in the soil. Air temperatures that permit plant growth normally range through some sixty-five degrees F. In general, the best range for plant growth is between fifty and ninety degrees F., though neither limit is firm. In tropical and subtropical areas the range is extended to higher than ninety-five degrees F. (recorded maximums are as high as 108 degrees F.), but in general, plant growth and the chemistry that sustains it tend to cease at air temperatures of about 105 degrees F.

Roots, as already suggested, coordinate and compromise growth temperatures. They have the mighty advantage of the modified temperatures that prevail in the soil around them. In undersoil levels as much as four to six feet below the surface they reach downward into everlasting springtime. This is a very special vantage for trees and other big, deep-rooted perennials. In areas nearer the surface, regardless of the air, soil temperatures rarely drop lower than a few degrees below freezing. In this protected medium, roots—particularly perennial roots—can and do practice their magnificent chemistry the year round. This, of course, is another very special resource of root sanctuary.

In ways that are usually easier to observe than to explain, the ability of roots to adapt themselves to existing soil temperatures is of tremendous importance to most vegetation, and pre-

sumably to all our major crops. This leads to due notice of the
fact that many crops require a certain temperature range for
effective flowering. For example, onions and rutabagas are
disposed to flower only at relatively low temperatures; peppers
usually flower only at relatively high temperatures. But in general the above-ground growth and development of vegetation
are known to be influenced profoundly by the degree of root
warmth. Thus, there is very sound reason to believe that healthier root structures and more abundant harvests depend more
on prevailing soil temperatures than on air temperatures.

One special area of distinctly brilliant progress is the temperature treatment of seeds prior to planting. Here a great deal
of excellent pioneering work has already been accomplished by
Russian, Western European, Australian, New Zealand, and American scientists, both together and apart. Currently, Soviet researchers seem to be taking the lead in profiting from thermal-
treatment control of seeds and other planting stock.

The basic premise is that many valuable crop plants, including the cereal grains, grow faster and better when the
temperature disparity between the root and the stem structure
is held to a minimum. In northern climates, the artificial cooling
of seeds is one means of accomplishing this. The mere act
of holding seeds for a while at or near the temperature of the
soil in which they are to be planted encourages more rapid
growth and reduces the time span between seeding and harvest.

In northern latitudes, where the growing season is extremely
short, seeds grow better if they are held at temperatures higher
than those that normally prevail in storage bins. In intermediate
climates, increase of growth rate is accomplished by reducing
the temperature of seeds for a period of eight to ten weeks
before planting. In the case of the internationally standard
wheat variety called Red Turkey, the accredited preplanting
treatment for northern zones includes storing the seeds for ten
weeks before planting at temperatures ranging from thirty-three

to thirty-seven degrees F., with moisture content no higher than sixty per cent.

On an average, the plants produced by temperature-treated seeds reach the grain-yielding stage 110 to 120 days after germination, which establishes a very important saving of approximately a month of growing time. In wheatlands such as those of the Ukraine, this seed treatment is said to increase the yield and improve the quality of the grain. Similar reports are made for oats and rye.

The essential chemistry involved in this treatment of seeds and their increased growth is still not well understood. Chemistry and genetics seem to be vitally interrelated here, since the seed embryo represents the climaxing attainment of root chemistry.

On many fronts—and often in areas of research that at first glance do not seem related—new light is being shed on the very special advantages of the root realm or rhizospheres. In Wisconsin recently I noted the work of another research team that keeps quite near to what some would list as scientific breakthroughs. John C. Walker, of the Wisconsin Agricultural Services, and Russell R. Larsen, a pathologist with the U. S. Department of Agriculture, are very ably exploring the threefold interrelationship of vegetative growth, genetics, and root-favoring soil temperatures.

One of their immediate and practical projects has been the development of a variety of cabbage that can withstand the fungi that cause root rot (clubroot). Since the 1920s this extremely antisocial root fungus has largely destroyed the commercial cabbage crop in what was once the American kraut belt. By turning to plant genetics, Walker and Larsen developed a cabbage that is highly resistant to the attacking root fungi. By continuation of selective breeding, the research team, as this is written, is in the labor of developing what its members term a cold-rooted strain, a variety of cabbage that will be able to endure the prevailing cool soil temperatures better than the rot-causing fungi can. This, in course, points up still another method of improving root health and strengthening the sanctuary of root spheres—by skillfully adapting crop species to the most prevalent soil temperatures.

Better understanding and use of prevailing soil temperatures are being demonstrated further by a variety of pioneering ventures that would transform some of the great grain grasses—particularly wheat, oats, and rye—to the status of winter-growing forage grasses. The key is persistent selective breeding to the end of producing roots hardy enough to stand lower-than-normal temperatures and the selective development of what can be termed winter-hardiness genes. As usual, this procedure involves more than theoretical, or even practical, genetics. Its base is root chemistry and genetics, or a prevailing shoptalk term is "chromosomatics"—meaning the capacity of winter-growing leaves to cling closer to the soil and thus take better advantage of the sun's radiation. This is of particular advantage during periods when both air and soil temperatures are lower than those ordinarily conducive to substantial growth.

The practical goal, in turn, is to provide green forage for livestock during the season when it is scarcest. Incidental benefits include reducing winter erosion, maintaining good soil texture, and keeping the countryside pleasantly green when most plants are dormant. The success of winter-growing cereal forages in great areas of the United States, Canada, Scandinavia, and Europe foretells a time when the temperate zones except in snow cover may be almost as consistently green as the tropics. In time, healthier root structures with their continuing root chemistry can almost certainly effect this green miracle. But there are still many other baffling and wonderful phenomena of roots that await alert and stubborn study.

The impressive successes in improving crop yields by means of very modest increases of root-sphere temperatures helps focus attention on more capable studies of soil-surface and root-level

climatology. Some call this promising newcomer microclimatology; none can deny its relevance to the fact that vegetation
lives and dies in a climatic realm that is mighty in influence
and usually quite tiny in reach. The thus-far-unmapped millions
and billions of Lilliputian weather zones are the ultimate life
realms of plants. It follows that the actual temperatures and
moisture conditions that prevail in the rhizospheres and biospheres and the air directly above them play a momentous part
in deciding the life and growth of vegetation. The readings,
like the decisive influences of this soil-level weather, can, and
usually do, vary impressively with immediate topography, alti-
tude, and wind directions, and within a given square yard or
even square foot. Certainly the "weather story" of the actual
living plants cannot be stated competently by prevailing weather
reporting, since vegetation frequently comes and goes by single
inches and degrees, not by regions or nations. When weather
changes, as it always does, the chemistry and the growth of
plants are accordingly influenced—with tremendous localization
and immediacy.

In the United States, microclimatology was rather literally
born during the summer of 1960, when an exceptionally observant member of the agricultural faculty of Cornell University
noticed an extremely puzzling inconsistency in the growth rate
of corn in the area of Ithaca, New York. In particular, Robert
M. Musgrave noted that even when moisture and sunshine conditions were uniform and favorable, the corn plants seemed to
grow faster on breezy days than on still days.

Musgrave could not explain the disparity on any ready, dependable agronomic basis. The only plausible explanation he
could come up with was that on still days the corn-growth
rate was being retarded by an inadequate supply of carbon
dioxide in the soil and the air. The young researcher was certain
that the atmosphere contained an abundance of CO_2; any
deficiency would have to be in the soil. Plant students are
fairly well agreed that a living plant produces only about 20

per cent of the carbon dioxide it requires. The rest is obtained from the outside air.

In his initial studies, Musgrave put transparent plastic cages around several corn plants to study what happened when they were cut off from airflow without a reduction of sunlight. Next, he devised some reflector meters to measure sun radiation on the growth surface of typical corn plants. He took daily growth measurements of plants in the cages and of others in the open. He also took humidity readings in the root spheres, the upper soil, at ground level, and at tassel tips. By these and other computations, he was able to confirm that during maximum growth the corn plant obtains up to 80 per cent of its carbon-dioxide supply as air "blow-in" and to deduce that the measurable growth reduction of the caged plants was related to a restriction of the carbon-dioxide supply, accompanied by reduced efficiency in the use of sunlight by the leaves. No less revealingly, the plants outside of the cages continued to grow more rapidly on breezy days than on still days.

Musgrave continued his study and measurement of air movement at ground levels. At this point, Robert Lemon, a physicist with the U. S. Department of Agriculture, joined forces with him in confirming the fact that surface air literally rolls in on a plant—a wheeling or turning motion at ground level.

Air roll brings to the plant carbon dioxide, nitrogen, and other airborne materials indispensable to its chemical functions and proper growth. Contrasts in the size and height of vegetation act to increase the turbulence of the ground-surface air. It follows that the intermixing of rows of short crops with rows of tall crops can serve to increase the movement of air at the soil surface. Helpful variants in crop placement are easy to make. Corn rows, for example, can be interspersed with rows of soy-beans or tomatoes; or sweet corn can be alternated with snap beans or common greens, and so on, to the end of increasing the roll-in of air. The researchers observed that hard barriers, such as dense hedges and walls, retard air roll and divert it upward. In general, a speed-up of air roll increases evaporation of moisture in the soil.

Conversely, the rate of air roll can be reduced by planting grasses, shrubs, trees, or row crops of fairly uniform height, and by keeping lawns or meadows uniformly mowed. In any case, availability of water to the plant is affected. Maximum access to carbon dioxide increases the ability of a plant to absorb water; slowing the air roll tends to diminish water absorption. Heat, of course, serves as an activator in these processes.

As researchers Musgrave and Lemon are among the first to confirm, the science of root-zone climatology is still in the borning and forming. Because roots are so momentously responsive to the entire range of meteorology, even the first dimly recognizable shapes in the new "discipline" significantly resemble a new key that may presently open a great new treasure trove of root-benefiting, people-saving knowledge. More and more related treasure troves are coming into view—for the most part directly beneath our feet.

At this place, the writer wants to insert a brief apology for not mentioning the scores and hundreds of other scientific findings and bornings that relate importantly or potentially to hardier and more provident roots and the ever-wonderful bounties therefrom. Dozens more that I have seen and studied, at least a hundred I know vaguely or by hearsay, and no doubt thousands that I don't know could and, in great part, should be properly included here. If this were six books instead of one, and if I were, say, forty years younger, they would be included.

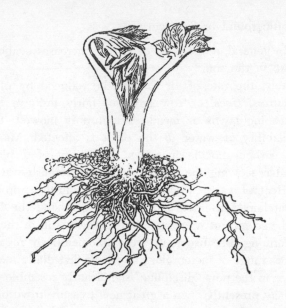

12

ROOTS OF TOMORROW

Man feeds mostly on vegetation, which is fed mostly by roots. This has been going on for a very long time.

For many root structures 100 years constitute hardly more than a passing day. For agriculture (from the Latin *ager* and *cultura* "field culture"), which now designates the whole wonderfully mixed-up science, art, trade, business, and folklore of growing crops and livestock on land, 100 years offer a fairly effective yardstick. At least since the time of Aristotle (384–322 B.C.), who was certainly no small potato of a farm commentator, at least once a century agriculture has tended to reset or reappraise its working patterns. There have been lapses during the ensuing twenty-three-plus centuries, and periods of more rapid changes, but for the most part, approximately once a century agriculture has changed sufficiently to be relabeled "new."

The foregoing observation is not original with this writer. About half a dozen decades before the beginning of the Christian era a more gifted farm reporter stated it as a general proposition and boldly predicted the likelihood of its continuation.

His name was Gaius Plinius Secundus, better remembered as Pliny the Elder, and he remains among the all-time champion country correspondents. This exceptionally gifted uncle of Pliny the Younger (himself no small scribbler) was deeply interested in roots and in duly protecting them from such deplorable mishaps as being denuded by avoidable erosion or crushed or gouged too soon after planting by overweight grazers such as oxen. Almost three centuries earlier, Aristotle was also giving deep thought to roots as provers of subsurface perpetuation of the "nondivorcement" of being and becoming, occurrence, and recurrence.

On that rather cosmic theme, it might be revealing fun to try taking a brief from-the-roots-up look at these United States of America 100 years from now. Without any bothersome claim of clairvoyance or abnormal prophetic powers, let's take a make-believe stroll through any real and reasonably typical American countryside a century from now, projecting from established facts or well-defined likelihoods to a given day in A.D. 2070.

What follows here is this writer's projection of what life, including roots, may be like a century, or slightly more, from now. The fact that the earth's population in A.D. 2070 is somewhere near 7 billion, instead of 8, 9 or even 10 billion as earlier demographers had feared and openly predicted, is another great testimonial and proof of the versatile providence of root spheres. The birth control hormone, which as early as the 1960s had begun proving its global capabilities for preventing or greatly reducing human conception, continues to prove its efficacy as a specific for population control. With due improvements, "The Pill" has steadily gained dependability and practical acceptance.

By, if not before A.D. 2070, people have grown generally aware that the source of the hormone compounds that are the basis of the effectiveness of "The Pill" is a once tropical and wild growing yam. The plant was first located and put in use in Mexico, principally in the jungle-strewn state of Oaxaca.

By 1968 the brilliantly competent Mexican Ministry of Agriculture had set up a special plant research station near Tuxtepec to conduct experiments in raising the yam as a cultivated commercial crop in an area where the wild-growing Dioscorea kin was believed to be indigenous. An official announcement of the opening of Mexico's then newest research center was first carried in dispatches by the Associated Press, the United Press-International and other major news services on February 14, 1968.

At that time practically all starting material for "The Pill" was being gathered and processed in Mexico, where most of the research and development work had actually taken place. It was then known that the yam (in 1968 still without an accredited botanical name) grew wild in the low wetlands of lower Mexico and that practically all of the earlier supply had been harvested and marketed by rural Indians.

First attempts to adapt the source plant to field or garden cultivation had not succeeded; the abrupt changeover to field culture in differing environments had produced yams with ineffectively small contents of the mysterious but highly effective hormone. But during 1968, the unimpressive tuberous root, which grows to the size and general appearance of a half-deflated football, was clearly responding to improved agronomy in its homeland area. The progress was already sufficient to give virtual assurance of an adequate international supply of the starting material. The generally hardy yam was destined to be an epochal factor in limiting the so-called population explosion supposedly set off by World War II.

No contribution by root spheres, or anything else, could alter the ready-accepted fact that millions of years had been required for planet Earth to acquire its first billion people and that barely two thirds of the twentieth century had been required for adding the second, third and part of the fourth billions, or that world population had approximately doubled again since the beginning of the final third of the twentieth century. But, at least, A.D. 2070 is demonstrating that root spheres continue to contribute more and more momentously to sustaining people as well as helping to restrict their rampant increase.

A.D. 2070 finds the United States one of the more sparsely peopled of the major powers. China, the newly-foaled West and East Asia Confederations, and the similarly young confederations of Africa, mainland Latin America, the Caribbean Islands and the South Pacific, are among the most densely populated countries. The United States, with barely a third of a billion people has come to typify the older nations with waning birthrates.

The reasons are self-evident: necessity for agricultural self-sufficiency; inability of the national economy to keep pace with even intermittent population explosions; and the stubborn inelasticity of the available employment in terms of job seekers at hand and clamoring for work. Generations earlier, the need for unskilled labor was virtually erased; by now the demand for semiskilled labor has melted away, and skilled labor and many specialized technicians are feeling the pinch of unemployment. For better or worse, now in 2070, the trials and tribulations of Social Security have grown to be an economic cross, and the so-called emergency relief for any major public has grown to be an antic that no mature nation dares attempt.

In A.D. 2070 the hurrah about opening "great and fertile" deserts to irrigation and the warming of arctic lands by means of atomic energy has pretty much quieted down insofar as both have already been attempted and attained, but with rather appallingly little increase of crop acreage in planting. Most, at least 85 per cent, of all tenable lands that required irrigation were already irrigated by 1970, and the enhanced competence of food production does not require or even recommend the excessively expensive increase of crop irrigation. No less notably, by 2070, these United States have grown to be a cautious country that would rather eat than fight, instead of vice versa, as in various decades of the primitive 1800s and the ridiculous 1900s.

To these thoughts while, or before, we stroll a century into the future might be added a few reflections in the realms of natural history? Long ago, in the now dim 1920s, perceptive scholars, beginning with the fact-loving, root-examining John Ernest Weaver, had learned and reported that plant ecology and human ecology are basically the same science. For both, living and growing roots are a prime and common denominator. This bears most directly on the fact that roots are ever more definitively the medium of man's survival with some degree of comfort and a vestige of dignity, and the ultimate deciders of land use.

Back in the mysteriously erratic 1930s, faced by a ruinous depression, the United States began to realize that vast areas of its farmland were no longer productive in terms of a rational, people-sustaining economy. Since it was too late to give a billion or more hopelessly inept or exhausted acres back to the Indians, we shook down the U. S. Treasury for "domestic allotments" and devised other schemes for paying farmers not to farm, or to help "resettle" those who could not be subsidized by means of such domestic allotment, then began to look for more durable ways of selling or giving a third of the nation back to the nation. By the 1950s, lucid Americans were beginning to realize that only the best—or at minimum, second-best—farmland could endure as such a bona fide, competitive agriculture. However, by—or, in some places, even before—the 1950s, important areas of the truly first-rate agricultural lands were already being "planted" with suburban and outer-suburban housing developments, factories, shopping centers, government buildings, and decentralized industries or were given over to highways, airport sites, public parks, forests, playgrounds, and wildlife refuges.

This continued to proliferate madly during the second half of the twentieth century.

By the closing 1960s, approximately 36,000 square miles of the United States, an area about the size of Indiana, had already been expropriated for highways and other roads. By 2070 the substantially increased roadway lands are approximately matched by airport and air-freight sites. By the last third of the twentieth century, land shortage in the United States, further intensified by the handing over of more than a billion acres to public domain, was being salved by the abatement of the human population explosion but without cogent heed of the tremendous birthrate increases in the Caribbean and nearby Latin-American countries, which could one day sweep over us like a ruinous tide. However, during the late twentieth century the United States had raised high and militant walls against immigration while seeking to sustain self-sufficiency in terms of food production.

One inevitable result is that in 2070 the American citizen feels cramped for space, only partly because of increased population, more because of the voracious increase of public domain. The land available for private use, including privately owned farms and other food-producing establishments is severely limited—a transcending challenge that has to be and is being met.

A century earlier, already overpopulated California, to name one roughly buffeted example, had pointed the way toward getting a great deal more food and fabrics (particularly cotton) from much fewer acres. Though largely a land of barren mountains and near deserts, California, between 1950 and 1970, had established itself as far and away the most versatile, skillful, and productive farming state in the nation—indeed, as one of the most impressive agricultural strongholds in the world. Despite appalling water shortages, erratic climate, and the merciless rape of its more fertile valley lands by sprawling new towns and cities, mushrooming suburbs, interminable shopping centers, industrial sites, highways and other roads, California had already

managed not only to hold and stabilize its place as the nation's
greatest farming state, but to steadily expand its leadership in
such food-producing areas as poultry, beef, dairy products, major
and minor fruits, and cotton. Even then, new and valuable plantings grew from soils that only a few years before had been
rated too poor even for grazing; even then, California cattle,
though relegated to browned mountains and foothills, already
included many of the finest beef animals and milk animals
that had ever walked the earth.

California's eye-opening agricultural success, by no accident
associated with some of the most able scholarship of roots then
obtaining in the United States or anywhere else, was also due
to a new type of farmer who set the pattern for the farmer
of the twenty-first century. A thoughtful man in a business suit,
he held a degree—often several degrees—from a state university
or college or accredited technical school. He was well schooled
in the basic and agronomic sciences, and he excelled in the use
and management of power machinery, including power pumps,
soil improvers, mobile sprays, and irrigation equipment. For good
measure, he had a working knowledge of civil and thermal
engineering, livestock husbandry, marketing, and accounting.

The "new" farmer selected his crops with painstaking care,
adapting them to the land available, and, having made his
choice, kept it on a perennial basis. He expected lean years
while crops were being established, but thereafter anticipated
continuing profits over many years, perhaps for a working
lifetime, conspiracies of realtors and/or politicians permitting. In
some instances his "specialty" involved placing two or even
three crops in a chosen routine of production. Whatever his
crop selection, root spheres and soils were his prime concern.
Where his predecessors had looked up, he looked down—and as
a producer of edible harvests, succeeded beyond the most exu-
berant hopes of the earlier pioneers, and that despite all handi-
caps and abuses imposed upon him.

But, to go back—or better say, forward—we are taking a stroll through a reasonably typical countryside on New Year's Day in 2070. By now, man has made substantial progress toward preventing the development of violent storms, and, except for snowfalls, the land is almost as green in January as it is in June (all the principal grasses are winter-green perennials).

Also, despite a rather formidable increase in motor vehicles—preponderantly battery-driven—town and country are quite convincingly wedded. For years the major cities have been sloughing away and most of them are being replaced by what are, in effect, city-states. The city-states are urban communities that hold in economic and cultural orbits their near and more distant suburbs and the outlying hinterlands. Often they include as many as twenty counties, or former counties. Though old in precedence and principle, the world trend is toward such units. As federal structures continued to gain and the old political units of township, county, state, and province lost competence and appropriateness, the city-state, certainly before A.D. 2000, became the prevailing body politic for serving the principal needs of the people.

In the city-state in which we are strolling we find a health center; a college; trading, youth, and federal centers; a theater; a library, an art gallery or working art center; a public market; and parks and playgrounds. All of these government services are obligatory. Obligatory also are transportation and storage facilities and accommodations for travelers, including local as well as interstate and international airports. The city-state serves also as a pumping and service center for pipelines, which are the fastest-growing and most efficient means of transportation for goods. Operated by atomically maintained pressures, these underground lines carry not only water, gas, and petroleum products, but also cements, bulk plastics, building materials, paints, sand, grain, flour, and a variety of packaged goods including parcel freight. Beyond the coastlines, transoceanic trade is now conducted principally by way of oversized submarines and high-speed, laminated "aquatic float missiles," which make up in speed—above 100 miles per hour—for their lack of size. The fifty-ton streamlined missiles carrying cargoes of a mere 100 tons or so cross the Atlantic in a single day with crews of only two or three men or women.

On the land, the pipelines free roads and streets of most of the heavy trucks, reducing surface congestion and smog. In many ways, strolling is more pleasant than it was a century ago. Rather evident to us is the new accent on trees and other plants in urban centers as well as in the suburbs and countryside. The greenery has a threefold purpose—to provide beauty, cleanse the air, and supplement the now standard "basic food issue," the federal ration of 2000 carefully computed calories a day for each person.

As indicated by those daubs of green that rise even from window boxes, growing part of one's own food to supplement government issues is both fashionable and necessary. The production, assignment, and distribution of basic rations have long been correlated with diet programing and in principal part assigned to qualified farmers in the nearest production zone. As of 2070, there are in the United States barely a million "primary" farms, managed by some two million licensed farmers (including many women), which are regularly producing about 80 per cent of the nation's food. A majority of the quarter-billion consumers help produce the remaining 20 per cent, plus some coveted extras.

As we stroll we find the little home gardens most attractive from both the visual and the culinary standpoint. We turn from the pedestrian walk of the main highway—now a pleasant country road which doubles as a residential street—and are impressed by the sidings of well-chosen grasses; the return of the grass age is very evident.

The grassy expanses are interrupted by some impediments, such as surface or temporary pipelines, and in many areas by miniature, open irrigation canals. There is little unused land, and fences are rare. Most of the houses are small, built of stucco or sheet aluminum, or economical alloys, or a combination of these materials, and the trend is toward the space- and money-saving round house. There are many small greenhouses, and almost every home has a vegetable garden and an attractive planting of berry bushes and dwarf fruit trees.

Indicated everywhere is the presence of underground structures—storage facilities, shops, shelters, and occasionally factories. People keep bobbing into the ground and out of it, like subway commuters. But this causes little distraction in the serenity of the ubiquitous grass, flowering shrubs, and well-tended earth. Little bare ground is visible; even clay banks and sand dunes are green with vegetation.

The predominant vegetation differs markedly in both appearance and fundamental physiology from that dimly remembered from the twentieth century. For the most part, foliage is much smaller and of much darker shades of green, including bluish greens and blackish greens. In the main, the root structures are substantially bigger than they were a century before, and very much hardier. Continuing progress with plant breeding is one of the many creative contributors to this ever-gaining bonanza; the work techniques range from expert selection of herbaceous parent strains to multicrossed hybrids which are being established with the help of elaborate electronic selectors, brilliantly synthesized growth stimulants and retardants, and budding and seeding expediters. These and other devices, in great part automated, serve to reduce error factors while strongly reducing the time lags that used to dog and impede even the best-planned efforts in vegetative genetics.

The goals of plant breeding, like the work techniques, have kept becoming more specific and sophisticated in terms of human diet needs. They include harvests with higher contents of protein, amino components, food minerals, vitamin complexes, and other people-benefiting nutrients. The improved delivery of food elements, particularly organic nitrogen, to the edible harvests rather than to discarded stalks and stover (in the grain crops, for example, the principal nitrogen production is no longer dissipated in the stems but is well concentrated in the seed or grain heads) has become a very notable hallmark of twenty-first-century agronomy. Its attainment has been accomplished primarily by the continued improvement of roots.

The measures of success still include the steadily increasing harvests. By now, however, the mere volume measures of food harvests—tons or bushels per planted acre—seem as ludicrous as they are obsolescent. To appraise volumewise, the bumper corn (maize) crop of the mid-twentieth century, around 100 bushels per acre, has long since been replaced by a minimal qualifier of 300 bushels, and the once boasted vegetable yields of 800 to 1000 bushels per acre are rated as wholly unacceptable now that 2000 bushels are commonly listed as minimal. But the decisive measure of harvest improvements isn't found in gross tons or bushels; it is found in greatly improved nutrition values. The grains, for example, now have digestible protein contents of from 8 to 12 per cent of edible bulk, which is closely similar to those of the twentieth-century high-protein foods such as lean beef (around 10 per cent protein), eggs (11 per cent), chicken (15 per cent), and so on.

There is no longer any extensive production of what used to be cheap mass foods, now that practically everybody realizes that the bulk carbohydrates are not and never were really inexpensive, and that, regardless of their local or world prices, and even as emergency sustainers of great numbers of people, the so-called cheap foods are not truly a bargain.

As the casual stroller may notice, there are still extensive plantings of the now deeply revered "reserve" grains—such as rye, which through the centuries has been one of the healthiest, best-rooted, and most efficiently harvestable of all the grains (big-seeded grasses), and oats, which since the ancient times of its emergence on the shorelands of the Mediterranean has excelled in terms of natural balance of nutrients. But even these age-old stalwarts have been conspicuously improved in the most logical manner—by way of their roots.

It follows that anybody with eyesight, even the most casual stroller, can see what "reasonably" healthy vegetation looks like. As yet completely (virtually 100 per cent) healthy vegetation remains the special grail of plant scientists, but the prevailing health levels have climbed to averages ranging from 70 to 80 per cent—in people-sustaining contrast to the previous century's tallies ranging from an estimated maximum of perhaps 55 per cent to God only knows how low.

The increasingly verdant landscapes also show grazing livestock, mainly dairy cows and beef cattle, and a few sheep. (Practically all poultry is grown indoors on commercial "battery farms.") Beef and dairy products still rank as coveted foods that even well-managed commercial farms cannot supply enough of to meet demand. A century ago, the ratio of cattle to people was one to one and eight tenths; a somewhat lower ratio now prevails.

However, the appearance of the cattle, still the most important domestic livestock, and their placement are very different. Except on federal land there are few large herds—not many of them more than of, say, 100 head. But now that the grasses are growing everywhere and have been so greatly improved in nutritive value and year-round availability, cattle and other grazing livestock can thrive and fatten on them without supplementary rations of grain.

The cattle we now see are not especially beautiful by twentieth-century standards. Except for the basic breeding stock, which is government owned and distributed (or, stated more formally, dispersed by artificial insemination), there are no more purebreds. For animals as well as for vegetation, this twenty-first is the century of the hybrid or selective cross. The cattle we see during our stroll are docile creatures with very small heads, thin legs, and narrow foreparts; the milking cows have huge udders, and the beef cattle have hindquarters which bulge with steaks and roasts.

Beside and behind the houses, and in community-owned pounds, one sees pigs. Swine remain one of the most efficient converters of waste materials into nutritious and appetizing meat, and meat has remained not only a prime desire of man but a well-established status symbol. The new-era pig is robust and roly-poly, fast growing, and mainly grass fed. Like the new cattle, it is varicolored, often with bizarre blots and splashes— an evident crossbreed. It is to be seen in town, suburb, and countryside. Zoning ordinances tend to agree that it is all right to keep a reasonably clean pig. Even in swankier areas, the pigpen is accepted as an asset. Pork has long been the secondary meat preference of mankind at large, and few would deny that home-raised pork is preferable to the deliberately inferior allocations to restaurants of cassava-soybean patties or those nasty little green algae salads.

Meanwhile many other meat equivalents are available, and some are actually more nutritious than real meat. The so-called analogue meats, made of vegetable mixtures fortified with strong protein concentrates from soy and other beans, peas, cotton seed, peanuts, palm fruits, etc., are being spun from plastic cordage and further benefited by a vastly improved chemistry of synthesized flavors and colors so that they look, taste, chew, and nourish like the "real thing" or possibly better.

The substitutions are very effectively supplemented by highly nutritive additives—amino- and mineral-rich grain mixtures designed to make any foodstuff sufficiently nutritious by sprinkling it with a few spoonfuls of the enricher, and featuring special concentrates, such as fish powder, with protein contents as high as 80 per cent. Advances in dehydration now make any principal dairy product, including whole milk, available to all, and the strategic combination of "superfreezing" by means of liquid nitrogen and vacuum drying or irradiation makes all principal foods storable at room temperatures. (They can be reconstituted merely by adding water.) Food drinks, meanwhile, have long been staples. The trade adages extol "pre-readied" foods and insist that no food is really produced until it is in the hungry consumer's mouth. In practically all countries, commercial food production is now mainly centered on producing edible roots and grains as the mainstay of the basic 2000-calorie diet. For the world at large, where hunger fighting is the principal warfare, the stronger-rooted perennial food crops, with varieties adapted to specific locations, are absolute necessities. Without these, no nation or confederation of nations can conceivably provide its people with even the minimal diet.

Though ours remains among the better-fed nations, with nutrition standards superior to those of the much deplored twentieth century, we are not exempted from the global food rules. That is why part, at least a third, of the red meats and most of the fresh fruits and vegetables are raised at home or in local community gardens.

Evidence of this new situation is everywhere. There are no vacant farms, very few idle fields, and no wastelands except for areas where the soil structure—no thanks to earlier Americans—has been irrevocably ravaged, or, at best, left depleted to a degree at which it can be brought back to producing only by centuries of colonizing by appropriate orders of mosses, primitive grasses, and the like. The forests, meanwhile, are practically all federally or state owned and maintained, and the same goes for most forest industries, from papermaking to the vastly improved manufacturing of building materials. Forest yields of lumber, fuel, and other products help to hold down the ever-menacing public debt. In general, the trees stand taller and thicker while the field and garden herbiage grows stubbier and generally smaller.

The observant stroller notices a significant use of public land. Practically every city-state has a botanic garden, planted exclusively with native flora. Its function is doubly educational. So many of the cultivated plants being grown come from other parts of the world that this may be the only place where the original vegetation can be seen. Such plantings are also of sound scientific value. By permitting indigenous or self-established vegetation of long standing to follow its natural course and reach its climax, naturalists and farmers can note which plants are best adapted to the area and correlate their crop plants with them. If the ultimate natural growth turns out to be grass or browsing vegetation, grains and forage crops are indicated. If trees come in, timber or orchards may merit special notice and experimentation.

But our glances keep returning to the home gardens, so ingeniously contrived. The windows of some homes are especially bayed to accommodate replacable plastic trays crowded with vigorously growing vegetables and flowers. Other plants are growing in small, inexpensive but efficient hothouses, provided with transparent, easily removed covers. Along paths and around small lawns are compact, highly productive fruit trees of impressive variety. In addition to apple, pear, peach, cherry, and plum trees, there are trees bearing citrus fruits, avocados, mangosteens, and other splendid fruits that were once associated only with the tropics. Continued genetic improvement, largely centered in root spheres, has adapted these to the climate of the temperate zone.

Probably, however, the perceptive stroller on New Year's Day in 2070 will be especially interested in the fresh vegetables that seem to be growing plentifully on a year-round basis in the miniature home greenhouses. Erect, well-pruned, and fruitful perennial tomato trees are to be seen; also sturdy, waist-high perennial bush beans crowded with crisp pods, and eggplant bushes heavy with purpling fruits.

Nobody sees anything remarkable in the sturdy, now well-proven perennial vegetables, nor in the long-commonplace in-

quiries about "which way to the tomato grove, the pea vineyards, the green-peppers station, or the perennial flower gardens
I helped plant forty years ago." Improved root strength is the
primary effecter of the change-over from annuals to perennials,
but as any plant historian or paleobotanist knows, this most
provident development is essentially a genetic revival—certainly
not an innovation. Consider, for instance, the tomato, which
continues to hold first place among all the above-ground vegetable harvests. For centuries, plant scholars have known that
the direct herbaceous (and South American) ancestor, the
tomatl, is a perennial herb (*Lycopersicon esculentum*) of the
mighty clan of nightshades. Similarly, the great capsicum or
food peppers clan, which at long last has gained acceptance
for its superb food values, is another of the provident nightshades that abound as perennial herbs and shrubs. Long ago,
certainly by the 1940s, expert appraisers, including eminent
nutritionists with M.I.T., were noting that Central American
sweet peppers, when self-returned to the wilds, on their own
power and rootage, reverted to the status of perennials, which
is ever indigenous and "natural."

Now, in 2070, the living accent is no longer on vegetarianism,
but on well-studied, ably developed food use of vegetables. The
living and eating background must include economic considera-
tions. As yet there is no need for converting or reverting all
the vegetable crops to perennials. Many of the root and tuber
vegetables are still quite adequate as annual crops, particularly
now that soil scientists and their companion geneticists have
succeeded in doing away with such yield and quality deter-
rents as "bearing siesta," or "late-season lag"—in the old times
the ever-expensive symptoms of root infection or other en-
feeblement. Even so, most of the melons and squashes are
now perennials; by standard vineyard culture and the "growing
waits" inherent in all annual crops. By contrast, white and sweet
potatoes, which even back in those stupid 1960s were averaging

around eight million digestible calories per acre, then three times the prevailing food crop averages, now (in A.D. 2070), even as annual crops, are yielding upwards of sixteen million calories per acre.

If given to reflection, now his most cherished luxury, the stroller will recall that while grain crops provided the foundation of the great civilizations of the past, vegetables contribute most to the food economy of the present. Historians of 2070 still enjoy recounting how home gardens "saved" Western Europe after World War II, how they inspired great improvement in crops through special concern for roots and, for good measure, did a great deal to help revive the war-wrecked European economy. Accompanying all this was a brilliant procession of genetic improvements, particularly of root vegetables. By 1960, European commercial gardening was concentrated in about 100,000 hectares or 247,000 acres in all, but included were some of the most abundant harvests ever recorded. The amateur's home garden was the prime motivation of a most historic sequence of crop and root improvement. By 1965, Holland alone had some 14,000 hectares—34,500 acres—in greenhouses or other transparent shelters.

Though few nations of A.D. 2070 are in position to rival the intensive efficiency of garden farming in the Netherlands, the observant stroller notices, even in the United States, a steadily improving economy of land use. There are no longer sprawling and untended fence rows. Private lands that adjoin forests are now preferred locations for many long-lived crops. So are stream edges and riverbanks. At long last, farmers and other land users are putting to valid use the unique advantages of "vegetative edges"—places where one plant environment gives way sharply to another. Thus, what used to be called topographical farming is now commonplace or standard.

The motivation for better land use remains tethered to the now global scarcity of land. By the middle of the twenty-first century, the total of earthly land stands at barely four acres per person, and despite imaginative and successful efforts at reclaiming and building up farmable land, the global total of land planted with food crops now averages less than a half acre per capita, certainly no more than 2420 square yards, or 21,780 square feet—somewhat less than half the prevailing average a century earlier.

This, of course, calls for the continuous improvement of practical farming and the many sciences and talents that it requires. Even though farming, as of 2070, is strongly supplemented by greatly improved commercial fishing and many improved sea-and-shore industries, and by synthetic recovery of atmospheric nitrogen for conversion to edible protein material, the primary responsibility for food production still rests with living roots in living soil plus the ever-advancing skills and creativeness of life scientists and, even more decisively, fewer but better farmers.

Back in the sixth decade of the twentieth century, the Food and Agricultural Organization of the United Nations reported that about 60 per cent of the earth's human population were still farmers. Even in that cruel and dimwitted era, students of food production insisted that the proportion of farmers was much too high and that too many farmers were producing too few pounds, kilos, or calories of food.

However, even during the 1960s trend-setting exceptions had begun appearing. In New Zealand, for example, which then led all nations in production of foodstuffs per man-hour of farm labor, one full-time professional farmer was producing enough food for himself and approximately sixty other people; in Australia the score was almost as high; in the United States, one for thirty-nine; in the United Kingdom, one for eight; in West Germany, one for six; both Japan and the U.S.S.R., one for two.*

* United Nations estimate based on average per-capita daily food intake of 2500 calories.

Between 1920 and 1965, food-crop production per man-hour of farm labor in the United States had increased 500 per cent, while average harvests per acre had gained about 70 per cent.* By 1965, too, production per man-hour of professional farm workers, again in the United States, was gaining more than twice as rapdily as industrial output. In order to meet minimum food needs of people, the passing of the century would find (so this writer guesses) the following prevailing tallies: New Zealand, one professional farmer feeds 100 people; the United States, eighty; Australia, eighty-five; Poland and Argentina, sixty a piece; Brazil, fifty; Germany, forty; the United Kingdom, thirty; the U.S.S.R., twenty-five; India, fifteen; Japan, ten; and China, five.

In no little part the gains in food production are gauged by reduction of wastes. Early in the twentieth century France's Ministry of Agriculture began compiling annual estimates of crop losses caused by plant diseases and normal harvest wastes. In 1910, although at that time French agriculture was outstandingly well advanced, the average harvest wastage was rather proudly reported as only 20 per cent. One after another, ministries or departments of agriculture of other nations began compiling and publishing estimated harvest losses caused by principal diseases of vegetation.

By the mid-1930s, some sixteen nations, including Germany, the United Kingdom, the Netherlands, Italy, Belgium, Czechoslovakia, Switzerland, Australia, and New Zealand, as well as France, had begun recording official estimates of crop losses to insects and other natural enemies. The ensuing estimates ranged from 16 to 29 per cent of the total harvests. During the quarter century between 1939 and 1965, the U. S. Department of Agriculture estimated average U.S. crop losses, in large part avoidable, by the following averages:

* Estimate of Bureau of Farm Statistics, U. S. Department of Agriculture, 1966.

CROP	Estimated losses by percentages of total harvests for the twenty-five-year period
Tobacco	24.9
Tomatoes (general or field crop)	24.5
Tomatoes (greenhouse or fancy)	8.4
Potatoes (white or Irish)	18.2
Cabbage	17.8
Cotton	14.5
Oats (grain)	11.6
Corn (field or grain)	8.6
Sweet corn	4.2
Apples	8.0
Sugar beets	7.2
Grapes	6.5
Wheat	6.5
Barley	5.5
Rye	1.3

Even back in 1939–69 it was common agreement that many other international crops—cacao, coffee, coconut, pineapple, pears, to cite but a few—were much less healthy than any of the fifteen listed above. In 1965, agricultural officials of Brazil and India estimated crop wastages, including harvest and storage losses, as approximately one third of the respective totals.

It had been fairly common agreement that without excellent work by pathologists, geneticists, crop-disease fighters, and others, especially vulnerable crops such as rice, tobacco, okra, cacao, pears, sugar beets, coconut, bananas, and citrus might have all but vanished from the earth. Cotton, potatoes, wheat, oats, alfalfa, clovers, and peanuts, left undefended, could almost certainly have been gravely reduced by diseases, chiefly of the root spheres. However, by the middle of the twentieth century the vegetative diseases were ceasing to occur in decimating, famine-producing epidemics so tragically recurrent in the nineteenth century. All this reiterated that man survives primarily because useful vegetation survives, and roots have most to do with that.

By 2070, every principal science is actively involved in the problem of crop improvement. Duly licensed farmers—the licensing of professional farmers is virtually global in practice—are directed to make extensive plantings only of crops with controlled hazards of disease losses; the permissible loss can be no more than 1 per cent of average gross yield. To help achieve this goal, the government owns, inspects, stores, and distributes all seeds, nursery plants, and other commercial planting stocks.

Still, various endemic infestations remain, and root diseases persist—preponderantly those caused by what were formerly called low-order pathogens (mostly bacteria, fungi, and viruses). Upgrading root health remains an unending task, but there is memorable, even historic progress to report. The two areas of most effective advances involve thermal treatment now attained by automated soil thermalizers (electrically powered heat applicators) and water fallowing. Both came into experimental use during the middle decades of the twentieth century; both are here to stay.

Flood fallowing—keeping croplands under water for periods ranging from several weeks or months to several years—usually requires government cooperation. For many soils and crops this water treatment is still the most efficient way of defeating some of the more serious fungus and virus diseases of roots. It is not effective for all types of root infection, but it destroys perhaps one third of the pathogens, including most of the more destructive ones. These include the air-breathing fungi, the *Fusarium* types, which leach and weaken root structures and open them to invasion by nematodes.

Quite inevitably, water fallowing requires improved management of water resources. Thus, though planet Earth had been growing warmer and drier for a very long time, the twentieth century had proved clearly that years of subnormal rainfall are not invariably years of crop loss or famine. Practical farming has long since proved that crops can be safeguarded during dry years by better (and minimized) cultivation, by improvement of the structure or tilth of the soil, by selective breeding and other genetic variations of crop plants, and most especially by better root development.

In the progress of root development it had long been known that a moderate scarcity of water, preferably a controllable scarcity, is often advantageous. When obliged to scramble for water, many root structures probe deeper and thereby become sturdier and more provident. Many prove themselves better able to endure dryness than are most of the root diseases that afflict them.

The early twenty-first century, like the late twentieth, had stressed the advantages to root health that are attained by reducing rather than increasing the quantity of irrigation water previously used. Controlled irrigation has long since proved the best and usually the least expensive crop insurance. Even in extremely dryland areas, successful irrigation can be accomplished by providing less than twelve inches of water a year. By 2070, the prevailing average irrigation for most of the basic crops, including edible roots, grains, grasses, and many orchard fruits, is consistently below twelve inches a year, nearer nine inches.

Food growers of the twenty-first century are distinctly aware that soil temperature is a vital factor in the productiveness of crop plants, one that profoundly influences the total community of soil lives. The progress in upgrading root health keeps stressing the development of strains and varieties that are precisely adapted to the prevailing temperatures of specific soil areas. The challenge is not new. Man has been facing it for hundreds of years, and nature presumably for hundreds of millions of years. One of the natural means, at which man now excels, is the development of hybrids.

Grains continue to gain most impressively by hybridization. This gained stature during the twentieth century, when hybrids made possible hugely increased corn harvests. Biennial varieties of corn had been developed by the end of the twentieth century and now, in 2070, perennial varieties are in widespread planting. Thus, the generous giant grass first known as maize now has roots that remain growing and otherwise active the year round, and with the coming of spring they revive with tremendous vigor, which presently crowds the stubby little stalks with clusters of heavy ears.

Oats and wheat, too, both probably of subtropical origin, have also been established, where desirable, as perennials long before 2070, or as year-round forage for livestock, or, as pioneered in Soviet Russia, as a perennial grain source. Perennial grain kaffirs (related to sorghum) and barley have gained in usefulness, while perennial rye, as already noted, has soared in importance; in much of the world rye now holds first place among the food grains. Here again, the change-over from annuals to perennials is effected primarily by determined improvement of the health and productive strength of root structures. By 2070, rice, too, has re-emerged as a perennial crop, and its capacity for feeding people is several times that of the now obsolete annual rice. Even so, this calorie-rich grain, like all above-ground crops, faces formidable competition from the edible roots.

The nutritious quality, combined with palatability, of edible roots and tubers is far and away the most impressive agricultural attainment of the twenty-first century. Progress here is still paced by a comparatively small list of time-proven root or tuber crops now greatly enriched in minerals and amino acids, the protein builders. These include taro (*Colocasia*), cassava (*Manihot*), the sugar beets (*Betas*) and related roots, the sweet potato (*Ipomoea*) and its various relatives, and the white potato (*Solanum*). In addition to these, some very important restorations and improvements of other food roots have been attained, and all are perennials. On the special-uses lists are such restored and improved entries as the biscuit root (*Lomatium*), the "wild sweet potato" (*Ipomoea pandurata*), arrowhead (*Sagittaria*) and its close relative, the tule root; bread root (*Psoralea*), and the bulbous sunflower formerly misnamed Jerusalem artichoke (*Helianthus tuberosus*). These, as noted, are planted as special-use crops, being nutritious, generally hardy, and in most instances well adapted to borderlands, desert edges, and forest margins, some even to stream beds.

The progressive development of edible roots crops stresses several worthwhile goals. Genetic procedures and a more fastidious selection of soils have given a better balance to the carbohydrates; minerals, oils, aminos, flavoring compounds, and even the chlorophyll of most crops have been augmented. To increase the chlorophyll, the upper portion of the root is encouraged to protrude above the soil. "Meat beets," meaning those with amino-protein contents of above, say, 3 per cent of gross weight, have long been commonplace.

In this late twenty-first century it is routine practice to combine natural and man-devised means for upgrading the nutritive value of root crops. Various procedures are used to reduce them to molasses or other essences, and aeration and skillfully controlled fermentation raise their content of organic nitrogens and other protein materials. After all, in a well-advanced atomic age the reshaping of molecular structures is always feasible if compatible starting materials are available in generous amounts, and no starting materials can be more compatible than extremely high-yielding edible roots.

The goal of more extensive and healthier root structures now applies almost uniformly to all food plants. Even in the nineteenth centuries the dependence of above-ground harvests on strong roots was clearly demonstrated; by the 1890s, in a few countries even earlier, nurserymen were successfully grafting or budding preferred varieties of fruit on the sturdier roots of kindred trees that were not capable of producing good fruit themselves.

Thus, many of the most cherished fruit crops began to grow from roots of a different variety of species. Hardy quince roots sustained many of the pit fruits and finer apples. By the early twentieth century, preferred citrus was growing from the hardier roots of low-value species, and English walnuts from the more durable roots of the common black walnut. Long before the twentieth century ended, plant scientists generally agreed that though a living plant must be viewed as unified protoplasm, its dependence upon its root structure is so great that root improvement is almost invariably the best means of crop improvement and hunger fighting.

Thus, in 2070, root lives are the subject of continued and concentrated study—laborious, complex but enormously rewarding. This study is the more abundant because of pioneer research. Early in the twentieth century, plant scientists were beginning to learn that when native vegetation is abruptly upset and replaced by a man-chosen single crop, the balance of soil life is drastically upset—and rarely for the better. Such practices as crop rotation upset the balance even further. In such instances the natural and more stable complex of soil microflora is replaced by a usually more simplified and, for that reason, less stable microflora. The long-endured processional of plowing, annual planting, fertilizing, biociding, insecticiding, and fungiciding with chemical compounds further conspired to impede the re-establishment of a well-balanced population of soil lives.

The more knowing soil scientists of the twentieth century had balked at charges that they were making matters too complicated. Matters *were* too complicated before they came along, and would remain too complicated. But fortunately, and not too late, scientists of the early twenty-first century had joined those of the late twentieth in affirming that when the lives of billions of people are at stake, legitimate scientific procedures must not be blocked arbitrarily by contriving groups. Regardless of the decade or century, food production is a serious business.

Realists by training and duress of need, the agronomists and life scientists of the twenty-first century continue to take a generally dim view of the so-called lower-life food sources. They are not entranced by the prospect of irradiated algae tanks, ocean plankton awaiting conversion into frozen slop, populous ant hills, worm farms, or grasshopper ranges. They know that the living soil, properly handled and thus tenanted by healthier roots of the higher plants, remains the most effective and satisfying medium for food production.

And, clinging like bacteria to provident roots, there is the knowledge that, as prime sustainers of green plants, roots are the very foundation of earthly life. Soil is the life-crowded film of earth that embodies survival; roots are the supremely competent chemical laboratories that deliver the material for survival. Healthy living roots, anchored in living soil, must be helped in the task of delivering more food as populations increase.

Scientists of the twenty-first century, carrying forward the research of their predecessors, are developing new planting materials, improved processing methods, and reduced waste involved in the distribution of produce. Diversification of crops is a general practice. In more and more instances, more than one portion of a food plant is being made edible; seeds or fruits are being supplemented by edible leaves, stems, buds, or even flowers. But the most impressive gains in food production are in the root spheres which by way of edible taproots, tubers, bulbs, rhizomes, or root crowns supply an ever-increasing proportion of man's food. Edible plants, like ball players, are being listed as double, triple, or quadruple threats or, more literally, capabilities. Varieties of crop plants that are exactly suited to specific soils and climates are now commonplace. Perennial crops already predominate; in time they will be practically unanimous, as in the very long ago.

Cultivation has been reduced to an absolute minimum. Forests and other lands of the public domain are kept consistently productive. As an absolute requirement for man's survival, water storages have been greatly increased, better controlled, and made more uniformly dependable. This is important to roots of plants, which at the same time are being made healthier, larger, and better adapted to the soils in which they grow. Despite greatly improved transportation, increasing percentages of food requirements are being grown within radii of a few miles, at most a very few hundred miles, of their principal consuming centers.

With each passing year the criterion of man's survival and well-being goes more and more decisively underground. For sustenance of the spirit or reflection on the Space Age, he may still look up, but for sustenance of the body, he looks down to the roots.

Bibliography

Chapter 1 MIRACLES BENEATH OUR FEET

BARDELL, ETHEL MARY, "Production of Root Hairs in Water," in *Botany*, Vol. I, No. 6. University of Washington, Seattle, 1941.

BERGMAN, H. F., "The Relation of Aeration to the Growth of Roots," a doctoral thesis. University of London, 1940.

BERKELEY, J., *The Development and Action of the Roots of Agricultural Plants at Various Stages of Their Growth*. Clowes, London, 1863.

CANNON, WILLIAM A., *Physical Features of Roots with Special Reference to Aeration of the Soil*. Carnegie Institution of Washington, D.C., 1925.

———, *The Root Habits of Desert Plants*. Carnegie Institution of Washington, D.C., 1911.

CLEMENTS, F. B., *The Role of Oxygen in Root Activity*. Carnegie Institution of Washington, D.C., 1921.

DRABBLE, ERIC, "On the Anatomy of Roots," Transactions of the Linnean Society of London, Series 2, *Botany*, Vol. VI, No. 3. London, 1912.

FIXTER, JOHN, *Growing Roots*. Dominion of Canada Agricultural and Forestry Commission, Ottawa, 1921.

GOWER, SAMUEL, "Alfalfa Root Studies," U. S. Department of Agriculture Bulletin 1087. Washington, D.C., 1959.

HILF, H. H., *Wurzelstudien an waldbaumen*. Sachafer, Hanover, 1927.

LUTZ, HAROLD J., "The Influences of Soil Profile Horizons on Root Distribution in White Pine," Yale University Forestry School Bulletin 44. New Haven, 1950.

MARKLE, MILLARD S., "Root Systems of Certain Desert Plants," a doctoral thesis. University of Chicago, 1925.

SANBORN, J. W., "Roots and Plants of Farm Crops," Utah Experiment Station Bulletin 22. Logan, 1959.

SNOW, LAELITIA M., *The Development of Root Hairs*. University of Chicago Press, 1905.

TEN EYCK, A. M., "A Study of the Root System of Cultivated Plants," Bulletin 43. North Dakota Agricultural Experiment Station, Fargo, 1934.

————, "A Study of the Root Systems of Wheat, Oats, Flax, Corn, Potatoes and Sugarbeets and of the Soils in Which They Grow," Bulletin 36. North Dakota Agricultural Experiment Station, Fargo, 1959.

WATERMAN, W. B., "Development of Root Systems Under Drought Conditions," a doctoral thesis. University of Chicago, 1939.

WEAVER, JOHN ERNEST, *The Ecological Relations of Roots.* Carnegie Institution of Washington, D.C., 1924.

————, *Root Development of Field Crops.* McGraw-Hill, New York, 1927.

————, *Root Development of Vegetable Crops.* McGraw-Hill, New York, 1927.

————, *Root Development in the Grassland Formation* (with W. E. Bruner). Carnegie Institution of Washington, D.C., 1922.

————, *Development and Activities of Roots of Crop Plants* (with F. E. Jean and J. W. Crust). McGraw-Hill, New York, 1927.

WILHELM, STEPHEN, "Diseases of Strawberries," Circular 494. Division of Agricultural Sciences, University of California, Berkeley, 1961.

INTERVIEWS: Dr. T. R. Hansberry, Associate Director, Shell Agricultural Laboratories, Modesto, California, 1965, 1966. Dr. Johannes van Overbeek, Modesto, California, 1965, 1966, 1967. Drs. William Snyder, Stephen Wilhelm, and Edward Baker, Division of Agricultural Sciences, University of California, Berkeley.

Chapter 2   ROOTS ARE ALSO FOR EATING

ABBOTT, CHARLES C., *Recent Rambles.* Lippincott, Philadelphia, 1933.

BIGGER, H. R., W. F. GRANARY, JOHN SQUAER, H. H. LANGSTON, and W. D. LESEUER, *The Worlds of Samuel de Champlain.* (Six volumes.) The Champlain Society, Toronto, 1922–1936.

BISHOP, MORRIS, *Champlain, the Life of Fortitude.* Knopf, New York, 1948.

BLANCHAN, NELTJE, *Nature's Garden.* Doubleday Doran, New York, 1929.

BONNOR, J., *Principles of Plant Physiology.* (with A. W. Galston). Freeman, San Francisco, 1952.

EMERSON, G. B., *Trees and Shrubs of Massachusetts.* Little, Brown, Boston, 1933.

HALL, HARVEY M., *Yosemite Flora.* Bureau of American Ethnology, Washington, D.C., 1940.

JEPSON, WILLIS LYNN, *Flowering Plants of California.* Jepson, San Francisco, 1941.

KAPHART, HORACE, (ed.), *Bailey's Standard Cyclopedia of Horticulture.* 1939.

MEDSGER, OLIVER PERRY, *Edible Wild Plants.* Macmillan, New York, 1959.

ROGERS, JULIA E., *Book of Useful Plants.* Doubleday Doran, New York, 1929.

SAUNDERS, CHARLES F., *Useful Wild Plants of the United States and Canada.* McBride, New York, 1935.

WILSON, CHARLES MORROW, *Meriwether Lewis of Lewis and Clark*. Crowell, New York, 1934.

————, *Samuel de Champlain: Wilderness Explorer*. Hawthorne, New York, 1963.

Chapter 3 A CLOSER LOOK

BALL, E., and C. E. T. MANN, "Studies in the Root and Shoot Development of the Strawberry," *Journal of Pomology and Horticultural Sciences*, Vols. V and VI. Burbank, 1958, 1959.

————, "The Origin, Development and Functions of the Roots of the Cultivated Strawberry," *Annals of Botany*, Vol. XLIV. Ithaca, 1961.

ESSAU, KATHERINE, "Vascular Differentiations in the Pear Root," *Hilgardia*, Vol. XV. University of California, Berkeley, 1961.

GOFF, E. S., "The Roots of the Strawberry Plant," transactions of the Wisconsin State Historical Society, Vol. XXVII, 1965.

HANSON, HERBERT C., "Comparison of Root and Top Development in Varieties of Strawberries." *American Journal of Botany*, Vol. XVIII. Berkeley, 1962.

JOHANSEN, D. A., *Plant Microtechnique*. McGraw-Hill, New York, 1940.

Jones, F. R., "Growth and Decay of Transient (Noncambial) Roots of Alfalfa," *Journal of American Society of Agronomy*, Vol. XXXV. Philadelphia, 1963.

NELSON, P. E., "Strawberry Root Anatomy with Special Reference to Black Root Rot" (with Stephen Wilhelm), *Phytopathology*, Vol. XLII, Berkeley, 1964.

RAWLINGS, T. E., *Phytopathological and Botanical Research Methods*. Wiley, New York, 1953.

————, *Techniques of Plant Histochemistry and Virology* (with W. N. Takahashi). National Press, Milbrae, California, 1952.

WEAVER, JOHN ERNEST, *Root Development of Vegetable Crops*. McGraw-Hill, New York, 1927.

U. S. DEPARTMENT OF AGRICULTURE, "Research in Plant Transpiration," Agricultural Research Service Report 70. Washington, D.C., 1962.

Chapter 4 ROOTS OF YESTERYEAR

AXLEROD, D. I., "Miocene Flora from the Western Border of the Mojave Desert" (with E. W. Berry). U. S. Geological Survey Paper 156. Washington, D.C., 1938.

BAYER, J. G., *Fern Allies*. Macmillan, London, 1887.

BOWER, F. O., *Origin of Land Flora*. Unwin Hill, London, 1948.

BROWN, R. W., "Green River Flora." U. S. Geological Survey Paper 156. Washington, D.C., 1939.

CHAMBERLAIN, C. J., *Gymnosperms, Structure and Evolution*. Unwin Hill, London, 1935.

CHANDLER, M. E. J., *Clay Flora*. British Museum, London, 1937.

CHRISTIE, M. E., "Conifers," *Bibliotheca Botanica,* Vol. XXVIII, Part 114. London, 1942.

DALIMORE, W., *A Handbook of Conifers* (with B. Jackson). Macmillan, London, 1917.

EAMES, A. J., *Morphology of Vascular Plants*. Wiley, New York, 1956.

ELIAS, M. K., *Tertiary Prairie Grasses*. Geological Society of America. Washington, D.C., 1939.

KIDSTON, R., *Old Red Sandstone Plants from Rhynie Chert* (with W. H. Lang). Unwin Hill, London, 1955.

MOHR, E. C. J., *The Soils of Equatorial Regions*. University of Michigan Press, Ann Arbor, 1962.

MORROW, CLARENCE, *The Organic Matter of the Soil*. University Printing, Minneapolis, 1919.

RIDDLE, G. H., *The "Minor Elements": Occurrence and Functions in Plant Life*. Riddle, New York, 1928.

ROBERTS, EDWARD, *Land Judging*. University of Oklahoma Press, Norman, 1960.

RUSSELL, E. J., *Soil Conditions and Plant Growth*. Lommany, New York, 1915.

SCOTT, D. H., *Extinct Plants and Problems of Evolution*. Macmillan, London, 1908.

SCARETH, G. D., *Man and His Earth*. Iowa State University Press, Ames, 1962.

SHUTT, FRANK T., "Western Prairie Soils of Canada," Canadian Ministry of Agriculture Bulletin 6. Ottawa, 1920.

STEPHENS, C., *A Manual of Australian Soils*. Commonwealth Scientific and Industrial Research Organization, Sydney, 1953.

STEWART, ROBERT, *Sulfur in Relation to Soil Fertility*. University of Illinois Press, Champaign, 1920.

TRIMBLE, HUGH, *Blades of Grass*. Australiana Society, Georgian House, Melbourne, 1949.

Chapter 5   SCHOLAR WITH PICK AND SHOVEL

WEAVER, JOHN ERNEST, Unpublished Papers, Microfilm Supplement 107. Carnegie Institution of Washington, D.C., 1956.

———, *Root Development of Field Crops*. McGraw-Hill, New York, 1927.

———, *The Ecological Relations of Roots*. Carnegie Institution of Washington, D.C., 1924.

———, *Root Development of Vegetable Crops*. McGraw-Hill, New York, 1927.

———, *Development and Activities of Roots of Crop Plants* (with F. E. Jean and J. W. Crust). McGraw-Hill, New York, 1927.

———, *Root Development in the Grassland Formations* (with W. E. Bruner). Carnegie Institution of Washington, D.C., 1924.

———, *Root Developments of Prairie Vegetation* (with W. E. Bruner). Carnegie Institution of Washington, D.C., 1924.

Chapter 6 THE TABLE OF ROOTS

BALFOUR, LADY EVELYN, *The Living Soil: Evidence of Importance to Human Health of Soil Vitality with Special Reference to Postwar Planning.* Faber & Faber, London, 1945.

BEAR, F. E., *Soil Is the Stuff of Life.* University of Oklahoma Press, Norman, 1962.

BLACK, C. A., *Soil Plant Relationship.* Wiley, New York, 1957.

BENNETT, HUGH H., *The Land We Defend.* Longmans, Green, New York, 1942.

COMBER, N. H., *An Introduction to the Scientific Study of Soils.* Arnold, London, 1960.

COOK, J. GORDON, *Our Living Soil.* Dial, New York, 1960.

CLARKE, G. R., *The Study of Soil in the Field.* Clarendon, Oxford, England, 1936.

DAUBENMIRE, R. F., *Plants and Environments.* Wiley, New York, 1947.

EVANS, MATHEW SEMPLE, *The Soil Runs Red.* Van Kampen, Chicago, 1948.

FAULKNER, EDWARD HUBERT, *A Second Look.* University of Oklahoma Press, Norman, 1958.

GOUGH, H., *Soil Insecticides.* Imperial Institute of Entomology, London, 1945.

HALL, SIR ALFRED DANIEL, *The Feeding of Crops and Stock: An Introduction to the Science of Feeding Plants and Animals.* Dutton, New York, 1911.

HEALD, F. D., *Manual of Plant Diseases.* Macmillan, New York, 1926.

HEWETT, E. R., *Good Land from Poor Soil.* Trenton Printing Co., Trenton, N.J., 1951.

HOLMES, JAMES M., *Soil Erosion in Australia and New Zealand.* Angus & Robertson, Sydney, 1946.

JACKS, GRAHAM V., *The Rape of the Earth: A World Survey of Soil Erosion.* Faber & Faber, London, 1939.

LURIA, S. E., *The Bacteriophage-Bacterium System.* Luria, London, 1953.

MITCHELL, ELYNE, *Soils and Civilization.* Angus & Robertson, Sydney, 1946.

TAMER, F. W., *The Microbiology of Foods.* University of Illinois Press, Champaign, 1954.

THAYSEN, A. C., *The Microbiology of Starch and Sugar* (with D. Galloway). Longmans, Green, London, 1930.

WAKSMAN, S. A., *Microbial Antagonisms and Antibiotic Substance.* Wiley, New York, 1945.

WILSON, G. S., *Principles of Bacteriology and Immunity* (with A. A. Milne). St. Martin's, London, 1944.

———, *Bacterial Viruses* (with W. W. C. Topley). Lippincott, Philadelphia, 1952.

Chapter 7 WANTED: HEALTHIER ROOTS

ACKERBERG, M., "The Application of Cytology to Herbage Plant Breeding," Imperial Agricultural Bureau Joint Publication 3. London, 1940.

ATWOOD, S., "Selecting Plants of Broomgrass for Ability to Grow at Controlled High Temperatures" (with R. A. MacDonald). *Journal of the Society of Agronomy*, Vol. XXXVIII. Ithaca, 1946.

AYRES, E., Energy Sources: *The Wealth of the World* (with C. Scarlott). McGraw-Hill, New York, 1952.

BAILEY, L. F., "Some Water Relations of Three Western Grasses." Welsh Plant Breeding Station (Aberystwyth) Imperial Bureau of Pasture and Forage Grasses, Series H. No. 12.

Bates, Marston, *The Prevalence of People*. Scribner's, New York, 1955.

BENNETT, M. K., *The World's Food*. Harper, New York, 1954.

BROWN, HARRISON, *The Challenge of Man's Future*. Viking, New York, 1954.

BURTON, G. W., "Quantitative Inheritance in Grasses," International Grassland Congress Report 6. U. S. Department of Agriculture, Washington, D.C., 1953.

DE CASTRO, J., *The Geography of Hunger*. Little, Brown, Boston, 1952.

DOWLING, J. W., "The Effects of Some Environmental Relations in Bulk Hybridization of Grass." *Journal of the American Society of Agronomy*, Vol. XXXV, 1946.

ELTON, C. S., *The Ecology of Invasions by Animals and Plants*. Methuen, London, 1959.

ELLIS, C. B., *Fresh Water from the Ocean*. Ronald, New York, 1954.

GEISE, A. C., *Cell Physiology Illustrated*. Saunders, Philadelphia, 1958.

JOHNSON, F. H., *Influence of Temperature on Biological Systems*. American Physiology Society, Washington, D.C., 1960.

LEAVITT, J., *The Hardiness of Plants*. Academic, New York, 1961.

MCKEE, J. E., "Looking Ahead for Water," *Engineering and Science*, Vol. XIX, No. 24. California Institute of Technology, 1951.

ORDWAY, S. H., JR., *Resources and the American Dream*. Ronald, New York, 1953.

OSBORNE, FAIRFIELD, *The Limits of the Earth*. Little, Brown, Boston, 1953.

REDDISH, G. F., *Antiseptics, Disinfectants, Fungicides, and Chemical and Physical Sterilization*. Lea & Febiger, Philadelphia, 1958.

Salter, R. M., *World Soil and Fertility Resources in Relation to Food Needs*. Chronica Botanica, Waltham, Massachusetts, 1948.

THOMAS, W. L., *Man's Role in Changing the Face of the Earth.* University of Chicago Press, 1956.

THOMPSON, SIR GEORGE, *The Foreseeable Future.* Cambridge University Press, 1955.

Chapter 8 BETTER BREEDING: BETTER ROOTS

BAKER, K. F., *Disease and Resistance of Major Floricultural Crops in California* (with P. H. Sciaroni). California State Floral Association, Los Angeles, 1952.

BELEHRADEK, J., *Protoplasma,* Monogram No. 8. Börntraeger, Berlin, 1964.

BLEDSOE, VISCOUNT C. B., *Report on Chemistry and Agriculture.* Second International Crop Protection Congress, London, 1949.

DOREY, O. G., *The Fruit Farm.* Faber & Faber, London, 1949.

EBELING, W., *Subtropical Entomology.* Lithotype, San Francisco, 1950.

FAY, C. R., *Great Britain from Adam Smith to the Present Day.* Longmans, Green, London, 1935.

FRASIER, SIR JOHN C., *The Golden Bough.* Macmillan, London, 1929.

———, *Spirit of the Corn and of the Wild.* Macmillan, London, 1936.

GOODOY, T., *Plant Parasitic Nematodes and the Diseases They Cause.* Dutton, New York, 1962.

HOWARD, L. O., *A History of Applied Entomology.* Smithsonian Institution of Washington, D.C., Vol. LXXXIV, Publication 3065, 1930.

LARGE, E. C., *The Advance of the Fungi.* Cape, London, 1941.

LEAVITT, J., *The Hardiness of Plants.* Academic, New York, 1961.

MARTIN, H., *The Scientific Principles of Plant Protection.* Arnold, London, 1960.

MOORE, W. C., *Report on Fungus, Bacterial and Other Diseases of Crops.* His Majesty's Stationery Office, London, 1948.

PARK, J. R., *Plant Breeding and Insect Damage.* Cape, London, 1964.

PLINY, the Elder, *Historia Naturales* (edited by H. Rackthorn). Harvard University Press, Cambridge, 1960.

ROBERTS, MICHAEL, *The Estate of Man.* Faber & Faber, London, 1951.

SYKES, G., *Disinfection and Sterilization.* Van Nostrand, Princeton, 1962.

WILLIAMS, C. V., *Horizons of Plant Genetics.* Unwin Hill, London, 1966.

Chapter 9 ROOTS AND LIGHT

ANDERSON, A., *Observations on Turnips.* Clowes, London, 1863.

BEAR, F. E., *Soil Is the Stuff of Life.* University of Oklahoma Press, Norman, 1962.

BLACK, C. A., *Soil Plant Relationship.* Wiley, New York, 1957.

BRYAN, A. E., *Comparison of Anatomical and Histological Differences Between Roots of Barley Grown in Aerated and Non-Aerated Soils.* Iowa State University Publication 1017. Ames, 1961.

CANNON, W. A., *Physiological Features of Roots with Special Reference to Their Relations to Aeration.* Carnegie Institution of Washington, D.C., 1922.

CLEMENTS, F. E., *Plant Ecology.* (with John Ernest Weaver). McGraw-Hill, New York, 1927.

DAVY, SIR HUMPHRY, *On Agricultural Chemistry.* The Queen's Stationer, Cork, Ireland, 1921.

DICKSON, J. G., *Diseases of Field Crops.* McGraw-Hill, New York, 1947.

HAMNER, K. C., *Light Sequences and the Biloxi Soybean.* Hamner, New York, 1958.

KISSELBACK, T. A., "Transpiration as a Factor in Crop Production," Nebraska Agricultural Experiment Station Research Bulletin 6. Lincoln, 1926.

LIEBIG, BARON JUSTUS VON, *Natural Laws of Husbandry.* Clowes, London, 1829.

LOWELL, W., "A Promising New Soil Amendment," *Science,* Vol. XCVII. London, 1951.

MCKINNEY, H. H., "High Temperature Venalization," *Botanical Review,* Vol. XVI. New York, 1950.

OOSTENBRINK, M., "Het Ardappelaaltje," *Versl. Med. Plant Dienst.* Amsterdam, 1959.

STOUR, R. H., *Growth and Survival in Soil.* University of Wisconsin Press, Madison, 1959.

TIMONIN, M. I., *Soil and Responses to Light.* (Reissue). Carnegie Institution of Washington, D.C., 1958.

WEAVER, JOHN ERNEST, *Root Development of Field Crops.* McGraw-Hill, New York, 1927.

———, *Root Development in the Grassland Formations.* Carnegie Institution of Washington, D.C., 1924.

Chapter 10 ROOTS AND CHEMISTRY

BELEHRADEK, J., *Temperature and Living Matter Protoplasms.* Börntraeger, Berlin, 1935.

BRODY, S., *Bioenergetics and Growth.* Reinhard. New York, 1945.

CLARKE, G. R., *The Study of Soil in the Field.* Clarendon, Oxford, 1936.

COOK, J. G., *Our Living Soil.* Dial, New York, 1960.

ELTON, C. S., *The Ecology of Invasion by Animals and Plants.* Methuen, London, 1958.

HALL, SIR ALFRED DANIEL, *The Feeding of Crops and Stock: An Introduction to the Science of Feeding Plants and Animals.* Dutton, New York, 1911.

HARDISON, J. R., "Seed Disorder of Forage Plants," U. S. Department of Agriculture Yearbook, Washington, D.C., 1953.

JOHNSON, F. H., *Influence of Temperature on Biological Systems.* American Physiological Society, Washington, D.C., 1960.

KOVA, V. A., "Soils and Natural Environments of China," U. S. Joint Publication Research Service No. 5967. Washington, D.C., 1960.

LEAR, BERT, "Soil Porosity as Related to the Distribution of Fumigant Materials" (with W. F. Mai, J. Feldmeir, and F. J. Speight). *Physiopathology*, Vol. XL, No. 17, New York, 1963.

NILLSON, F., "Herbage Chemistry and Plant Breeding in Sweden," Imperial Agricultural Bureau, Joint Publication 3. Stockholm, 1960.

ORDISH, F. G., "A Hundred Years of Lime Sulphur," *Agriculture*. London, June, 1951.

PUTNAM, P., *Energy of the Future*. Van Nostrand, Princeton, 1953.

RAWLINGS, T. E., *Phytopathological and Botanical Research Methods*. Wiley, New York, 1933.

————, *Techniques of Plant Histochemistry and Virology* (with W. N. Takahashi). National, Milbrae, California, 1952.

REDDISH, G. F., *Antiseptics, Disinfectants, Fungicides, and Chemical and Physical Sterilization*. Lea & Febiger, Philadelphia, 1958.

REDNICK, W. M., *Chemistry of the Living Root*. Howley & Cane, London, 1966.

WEAVER, JOHN ERNEST, *The Ecological Relations of Roots*. Carnegie Institution of Washington, D.C., 1924.

Chapter 11 ROOTS: BATTLEGROUND OR SANCTUARY

ALLEN, L. H., *Radiant Energy Exchange Within a Corn Canopy* (with C. S. Yocum, and E. R. Lemon). U. S. Department of Agriculture, Ithaca, 1963 and 1966.

BUSINGER, J. A., *Cold Protection with Irrigation*. University of Washington, Seattle, 1963.

CHASE, J. E., and associates, *Principles of Virology*. University of Florida Press, Gainesville, 1967.

FAULKNER, EDWARD HUBERT, *Plowman's Folly*. University of Oklahoma Press, Norman, 1943.

FEICHTMEIR, E. F., *A Study of Vegetative Procedures for Soil Building*. Colorado Experiment Station, Boulder, 1965.

FRITZCHEN, LEO J., *Microclimatology Data Handling* (with C. H. Van Buell). Agricultural Research Service, Tempe, Arizona, 1963.

GRIFFITHS, J. H., *Some Climatological Patterns in the Tropics*. Department of Oceanography and Meteorology, Texas A. & M. College, College Station, 1965.

HENDERSHOT, C. H., *The Use of Chemicals for Increasing Cold Resistance in Citrus*. Citrus Experiment Station, Lake Alfred, Florida, 1963.

JOOS, L. A., "Threshold Freeze Data Statistics," U. S. Weather Bureau, Washington, D.C., Fifth National Conference on Agricultural Meteorology, Lakeland, Florida, 1963.

PALMER, W. C., *Further Investigation of Micro-Meteorology*. U. S. Weather Bureau, Washington, D.C., 1966.

REYNOLDS, B. L., *John Deere*. Lawrence, Racine, Wisconsin, 1961.

SCARPA, M. J., "The Use of Dew Points in Plant Disease Forecasting" (with L. C. Ranier), U. S. Weather Bureau Special Report 110. Washington, D.C.

U. S. DEPARTMENT OF AGRICULTURE, "Shortage of Carbon Dioxide on Sunny Days." Washington, D.C., January 25, 1961.

Chapter 12 ROOTS OF TOMORROW

AYRES, E., *Energy Sources, the Wealth of the World* (with C. Scarlott). McGraw-Hill, New York, 1952.

BATES, MARSTON, *The Prevalence of People*. Scribner's, New York, 1955.

BEAR, F. E., *Soil Is the Stuff of Life*. University of Oklahoma Press, Norman, 1962.

BENNETT, H. H., *The Land We Defend*. Longmans, Green, New York, 1942.

BENNETT, M. K., *The World's Food*. Harper, New York, 1954.

BONAR, J., *Darwin and His World*. Allen & Unwin, London, 1924.

BRANDWEIN, P. F., *The Gifted Student as a Future Scientist*. Van Nostrand, Princeton, 1949.

BRODY, S., *Bioenergetics and Growth*. Reinhold, New York, 1945.

BROWN, HARRISON, *The Challenge of Man's Future*. Viking, New York, 1954.

CLARKE, G. R., *The Study of Soil in the Field*. Clarendon, Oxford, 1936.

COMBER, NORMAN M., *An Introduction to Soil Science*. Arnold, London, 1960.

COOK, J. G., *Our Living Soil*. Dial, New York, 1960.

DAUBENMIRE, R. F., *Plants and Environments*. Wiley, New York, 1947.

ELLIS, C. B., *Fresh Water from the Ocean*. Ronald, New York, 1954.

FOOD AND AGRICULTURE ORGANIZATION, UNITED NATIONS, *World Conditions and Prospects*. New York, 1949.

————, *State of Food and Agriculture*. New York, 1953, 1954, 1955, 1959, 1964, 1966.

HEWITT, E. R., *Good Land from Poor Soil*. Trenton Printing Co., Trenton, 1951.

JACKS, G. V., *The Rape of the Earth: A World Survey of Soil Erosion*. Faber & Faber, London, 1961.

MITCHELL, ELYNE, *Soils and Civilization*. Angus & Robertson, Sydney, 1946.

ORDWAY, S. H., JR., *Resources and the American Dream*. Ronald, New York, 1954.

OSBORNE, FAIRFIELD, *The Limits of the Earth*. Little, Brown, Boston, 1953.

RIDDLE, G. H., *The "Minor Elements": Occurrence and Function in Plant Life* (revised). Riddle, New York, 1958.

ROBERTS, E., *Land Judging*. University of Oklahoma Press, Norman, 1960.

RUSSELL, SIR E. J., *World Population and Food Supplies*. Allen & Unwin, London, 1954.

SCARETH, G. D., *Man and His Earth*. Iowa State University Press, Ames, 1963.

STAMP, L. D., *Land for Tomorrow*. Indiana University Press, Bloomington, 1952.

STEPHENS, C., *A Manual of Australian Soils*. Commonwealth Scientific and Industrial Research Organization, Sydney, 1953.

THOMAS, WILLIAM L., *Man's Role in Changing the Face of the Earth*. Cambridge, England, 1955.

Trimble, Hugh, *Blades of Grass*. Australiana Society, Georgian House, Melbourne, 1949.

THOMPSON, SIR GEORGE, *The Foreseeable Future*. Cambridge University Press, 1955.

VOGT, WILLIAM, *The Road to Survival*. Sloane, New York, 1948.

WILSON, CHARLES MORROW, *New Crops for the New World*, Macmillan, New York, 1945.

WEYTINSKY, W. S., *World Population and Production* (with E. S. Weytinsky). The Twentieth Century Fund, New York, 1953.

Index